52

THE

INTERNET:

BRAVE NEW WORLD?

Institute of Ideas
Expanding the Boundaries of Public Debate

DEBATING MATTERS

THE
INTERNET:
BRAVE NEW WORLD?

Institute of Ideas
Expanding the Boundaries of Public Debate

Dolan Cummings
Peter Watts
Chris Evans
Ruth Dixon
Helene Guldberg
Sandy Starr

Hodder & Stoughton
A MEMBER OF THE HODDER HEADLINE GROUP

303.4833 WAT
BKNCRM

DEBATING MATTE

Orders: please contact Bookpoint Ltd, 130 Milton Park, Abingdon, Oxon OX14
4SB. Telephone: (44) 01235 827720. Fax: (44) 01235 400454.
Lines are open from 9.00–6.00, Monday to Saturday, with a 24-hour message
answering service. Email address: orders@bookpoint.co.uk

British Library Cataloguing in Publication Data
A catalogue record for this title is available from
the British Library

ISBN 0 340 84841 3

First published 2002
Impression number   10 9 8 7 6 5 4 3 2 1
Year   2007 2006 2005 2004 2003 2002

Typeset by Transet Limited, Coventry, England
Printed in Great Britain for Hodder & Stoughton Educational, a division of
Hodder Headline Plc, 338 Euston Road, London NW1 3BH by Cox & Wyman,
Reading, Berks.

DEBATING MATTERS

CONTENTS

PREFACE

Since the summer of 2000, the Institute of Ideas (IoI) has organized a wide range of live debates, conferences and salons on issues of the day. The success of these events indicates a thirst for intelligent debate that goes beyond the headline or the sound-bite. The IoI was delighted to be approached by Hodder & Stoughton, with a proposal for a set of books modelled on this kind of debate. The *Debating Matters* series is the result and reflects the Institute's commitment to opening up discussions on issues which are often talked about in the public realm, but rarely interrogated outside academia, government committee or specialist milieu. Each book comprises a set of essays, which address one of four themes: law, science, society and the arts and media.

Our aim is to avoid approaching questions in too black and white a way. Instead, in each book, essayists will give voice to the various sides of the debate on contentious contemporary issues, in a readable style. Sometimes approaches will overlap, but from different perspectives and some contributors may not take a 'for or against' stance, but simply present the evidence dispassionately.

Debating Matters dwells on key issues that have emerged as concerns over the last few years, but which represent more than short-lived fads. For example, anxieties about the problem of 'designer babies', discussed in one book in this series, have risen over the past decade. But further scientific developments in reproductive technology, accompanied by a widespread cultural distrust of the implications of these developments, means the debate about 'designer babies' is set to continue. Similarly, preoccupations with the weather may hit the news at times of flooding or extreme weather conditions, but the underlying concern about global warming and the idea that man's intervention into nature is causing the world harm, addressed in another book in the *Debating Matters* series, is an enduring theme in contemporary culture.

At the heart of the series is the recognition that in today's culture, debate is too frequently sidelined. So-called political correctness has ruled out too many issues as inappropriate for debate. The oft noted 'dumbing down' of culture and education has taken its toll on intelligent and challenging public discussion. In the House of Commons, and in politics more generally, exchanges of views are downgraded in favour of consensus and arguments over matters of principle are a rarity. In our universities, current relativist orthodoxy celebrates all views as equal as though there are no arguments to win. Whatever the cause, many in academia bemoan the loss of the vibrant contestation and robust refutation of ideas in seminars, lecture halls and research papers. Trends in the media have led to more 'reality TV', than TV debates about real issues and newspapers favour the personal column rather than the extended polemical essay. All these trends and more have had a chilling effect on debate.

But for society in general, and for individuals within it, the need for a robust intellectual approach to major issues of our day is essential. The *Debating Matters* series is one contribution to encouraging contest about ideas, so vital if we are to understand the world and play a part in shaping its future. You may not agree with all the essays in the *Debating Matters* series and you may not find all your questions answered or all your intellectual curiosity sated, but we hope you will find the essays stimulating, thought provoking and a spur to carrying on the debate long after you have closed the book.

Claire Fox, Director, Institute of Ideas

NOTES ON THE CONTRIBUTORS

Dolan Cummings works at the Institute of Ideas, and is commissioning editor for the arts and media section of the *Debating Matters* series. He is the author of *In Search of Sesame Street: Policing Civility for the 21st Century* (1999) and edits the Institute of Ideas website (www.instituteofideas.com).

Ruth Dixon began her career in administration, working as deputy director of an independent housing association. She taught business languages on a part-time basis for several years while her children were small and then moved back into full-time employment as an administrator and then as deputy manager with a journalism training company. Ruth was responsible for running and developing the Internet Watch Foundation hotline service from its launch in December 1996 until March 2002. In November 1999 she was elected as the first President of the INHOPE Association, which brings together internet hotlines from across Europe.

Chris Evans is lecturer in multimedia computing in the School of Business and Management at London's Brunel University. He carries out research into computer interface design and is responsible for pioneering work on virtual learning. Chris has been a champion of freedom of speech on the internet since its inception and is the founder of one of the leading cyber-liberties organizations, Internet Freedom (www.netfreedom.org). He is also a regular columnist for *Practical Internet magazine*. Chris can be contacted on chris@netfreedom.org.

Helene Guldberg and **Sandy Starr** are Managing Editor and Press and PR Officer respectively at the online publication *spiked* (www.spiked-online.com). Based in London, with a global outlook, *spiked*'s ambition is to champion unorthodox, enlightened thinking and break new ground in online journalism. *spiked* stands for freedom of expression on the web, and aims to be 'online, off-message'. Guldberg was formerly co-publisher of *LM magazine*, while Starr was founding editor of the online magazine *Culture Wars*.

Peter Watts is a sociologist with a particular interest in the sociologies of science and medicine. After completing his PhD at Brunel University, he worked as a research fellow at Warwick Business School, on a project studying public access to the internet and social inclusion as part of the Economic and Social Research Council's recent 'Virtual Society?' research programme. He is now a lecturer in sociology at Canterbury Christ Church University College.

INTRODUCTION
Dolan Cummings

O, wonder!
How many goodly creatures are there here!
How beauteous mankind is! O brave new world,
That has such people in't!

These could be, but are not, the words of somebody who has just discovered the internet. In fact, they come from Shakespeare's play, *The Tempest* (Act 5, Scene 1). They are spoken by Miranda, a girl who has spent her life on a remote island and has never set eyes on a single human being except her father. When a ship is wrecked on the island and its motley crew stumbles onto the shore, she cannot contain her excitement and joy. Anyone encountering for the first time the wealth of human activity on the net must feel a similar elation.

The phrase 'brave new world' was more famously taken by Aldous Huxley as the title for his novel (published in 1932) about a high-tech and apparently happy society which is in fact deeply authoritarian and sinister. Miranda's naively enthusiastic phrase is now a byword for the novel's dystopian vision, pronounced sardonically by people wary of the promises of science and technology and of human enterprise more broadly. The internet is the latest technology to be greeted with a combination of Miranda-like optimism and Huxleyesque pessimism.

At the extremes, the internet is seen as a revolutionary development that is transforming the economy and society and even throwing into question what it means to be human. For some, the prospect of living in a 'virtual society' is an exciting one and the sooner we leave behind the boring old world of flesh, the better. Conversely, some regard the internet as a menace, threatening to reduce us all to passive consumers incapable of forming real relationships with real people.

For most people however, all this is a bit much. Few of us really believe that the internet is transforming society as comprehensively as it is claimed on the fringes. Nonetheless, the internet is undoubtedly changing the way we live and work, and in several areas, these changes are controversial. The internet has had a dramatic effect on discussions about free speech, business and democracy.

Because the internet is essentially a means of publishing and distributing information, inevitably it has stimulated controversy about free speech and censorship. The fact that the net is not owned or controlled by any single authority means that it is virtually impossible to regulate fully. For most, this liberation is something to be celebrated. In particular, it is pointed out that people living under oppressive governments now have greater freedom to distribute information and alternative political ideas. At the same time, however, many worry about the unchecked distribution of pornography and other offensive material such as racist and other 'hate speech'.

Certainly there is a massive amount of pornography on the internet, ranging from the merely titillating to the truly disturbing. And on the internet any bigot is able to publish views that are offensive and even incendiary. This is the flipside of the greater freedom afforded by the medium.

More worrying still, perhaps, is the internet's potential as a discussion forum for international terrorists. If the technology can be used by activists who want to challenge tyrannical regimes, it can also be used by those who pose a threat to democratic societies. Following the devastating attacks on America in September 2001, there was a great deal of speculation about whether the terrorists might have used the internet to plan their activities. Indeed, the US Government subsequently introduced legislation to give the police more powers to monitor private communications on the internet.

Ironically, this kind of surveillance is another concern. Information technology makes it easy to gather and store information about people, often without their knowledge or consent. Companies, government agencies and even individuals could potentially use the technology to snoop on internet users. The more dependent we are on electronic communication and data storage, the more vulnerable we are to such intrusion. The controversy is exacerbated by the ongoing development of encryption software, which allows internet users to protect their privacy. This lets people feel more secure about using electronic communications themselves, while heightening fears about unmonitored criminal activity.

The second area in which the internet has become a hot topic is the business world. Many commentators have talked about a whole 'new economy' based on high-tech information and communications technology. There are some high profile successes such as Amazon, the online bookseller that has become a household name. In strictly economic terms, however, the story is less happy. The 'dotcom' boom of the 1990s was short lived, and many 'new economy' firms did not make it into the new millennium.

Despite this setback, however, there are still high hopes that the internet will provide the focus for a new kind of post-industrial, and even post-material economy based on culture, entertainment and ideas. It remains to be seen, though, whether the internet will continue to be used primarily as an adjunct to existing economic activity – as a cheap means of selling goods and so on – or whether it really will transform the way we produce and, indeed, what we produce.

One of the greatest claims made for the internet is that it can revive democracy. There are two forms of this idea. The first is technical: for example, people could vote online. The internet could make it easier and more convenient to participate in existing democratic structures and perhaps reverse declining voter turnout. This seems sensible enough, except that it is generally agreed that the problems facing democracy run deeper than the inconvenience of getting to a polling station. Similarly, websites set up by local authorities and members of parliament make it easier to get information about what is going on in politics, but these are unlikely to inspire people who have little interest in politics or faith in its institutions to begin with.

A more thoughtful model of online democracy has the internet as a site for people to come together in new ways, to discuss new ideas. Certainly the net has been used extensively by various campaigning organizations, especially the current 'anti-capitalist' movement. Again, there is a technical aspect to this: leaflets and petitions can be sent to thousands of people across the globe in seconds, while multimedia websites can facilitate discussion and document activities. And the technology is increasingly cheap, especially compared with conventional publishing. More than this though, the internet is sometimes seen as democratic in a third sense. Anybody can post material on the net without having to get it past an editor or other gatekeeper.

People who are unable to get articles published in newspapers or magazines can publish on the web for a potentially global readership. This is certainly a good thing for those trying to promote new political and other ideas, as well as frustrated authors. The disadvantage is that readers have no guarantee of quality. It is often noted that there is a great deal of rubbish on the net and that much of the information available is unreliable, out of date or downright misleading. The same is certainly not true of the essays in this collection, each of which addresses one or all of the issues just discussed from a different perspective.

Dr Peter Watts is a lecturer in sociology at Canterbury Christ Church University College. He was previously member of a research team at Warwick Business School studying public access to the internet and social inclusion, as part of the Economic and Social Research Council's 'Virtual Society?' programme. Watts is sceptical of the idea that the internet really is bringing about a fundamental transformation of society. He cites research that indicates people tend to use internet in ways that suit their already established lifestyles, rather than adapting themselves to the medium. From this perspective, talk of a new, virtual society tells us more about the kind of society we already live in than about the potential of new technology.

One thing that has certainly been dramatically affected by the internet is copyright. The internet has made it very easy to copy and distribute books and music, making a mockery of existing copyright laws designed to ensure writers and artists receive payment for their work. Dr Chris Evans, a lecturer in multimedia computing in the School of Business and Management at London's Brunel University, examines the idea of intellectual property. As the founder of Internet Freedom, Evans is concerned that attempts to protect copyright online will undermine our ability to make the most of the medium.

Evans looks at the history of copyright and at its application online, arguing that the original rationale has been hijacked by publishing companies more concerned with protecting their own profits than with supporting a vibrant intellectual culture.

Ruth Dixon, former Deputy Chief Executive of the Internet Watch Foundation (IWF), examines some of the challenges presented by the internet and suggests ways these can be dealt with. She acknowledges the great benefits of the technology, but points out that the internet is sometimes abused. For Dixon, an extreme libertarian approach would end up depriving many people of the full potential of the internet, because they would learn to shun the medium altogether for fear of being exposed to offensive material. The IWF has developed a strategy of 'self-regulation', whereby internet service providers are encouraged to remove illegal or potentially illegal material and the IWF itself monitors the net in order to inform them where such material appears.

Helene Guldberg and Sandy Starr, of the online publication *spiked*, make an appraisal of the real potential of information technology and the net. They look at the two conflicting responses to the technology, labelling these technophobia and technophilia, and argue that both make the same mistake in attributing the internet with powers of its own, rather than seeing it as a tool that can be used in various ways. Guldberg and Starr see this mistake as the main barrier to developing the technology, arguing that instead of expecting the net to solve their problems, businesses need to think about how best to use it. Similarly, they argue that there is nothing inherently libertarian about the internet and that both free speech and privacy must be rigorously defended, even from the kind of self-regulation advocated by Ruth Dixon.

One final but important concern about the internet is more spiritual. Although the net is used primarily for communication, people using it are typically sitting alone in front of a computer screen. Some feel that this kind of 'virtual' communication threatens to undermine real human contact and relationships. I hope that this book will counter any trend in this direction by stimulating public discussion and debate both online and in the 'real' world.

Essay One

CYBER SCHMYBER: OR HOW I LEARNED TO STOP WORRYING AND TO LOVE THE NET

Peter Watts

Perhaps the first thing to engage with when trying to address the question 'The internet: Brave new world?' is to consider why – or indeed whether – we should be asking that question at all. I admit that at first sight this question appears to be a reasonable starting point for enquiring into the momentous technical, economic and cultural event that is 'the internet' and, anyway, it is only a title. Nevertheless, I have my doubts about it. I do not mean, of course, that nothing significant is happening. On the contrary, recent years have seen a rapid and far-reaching expansion of the internet and its applications, with more and more people getting online. Moreover, it is doubtless that this seemingly relentless growth of the internet will engender a multitude of social, political and economic impacts, both domestic and global, which must urgently be enquired into and analysed and understood. However, before embarking on such a project, it is important to think about how the central analytical questions should be framed, because the way in which a question is framed establishes the terms of reference for the subsequent discussion and hence shapes the conclusions finally drawn. Analytical terms are never neutral, but always implicitly reflect certain histories and worldviews, politics and ideologies (S. P. Wilbur, 'An archaeology of cyberspaces: Virtuality, community, identity', in D. Porter (ed.), *Internet Culture*, 1997).

This is why I chose to start this essay by querying the title of this book. It seems to me that the phrase 'brave new world' reflects a particular way of imagining the internet as a social problem. On the one hand, it is possible to take this phrase in a literal sense, albeit within its general cultural context. Read in this way, it speaks of the exciting possibilities of new computer-mediated frontiers, of new life and new civilizations, of boldly going and so on. On the other hand, the phrase may alternatively (and/or simultaneously) be taken as a reference to Huxley's novel. In this case, it must be understood ironically, a connection being made between the internet and the author's dystopic vision of a technological future.

The point, however, is that, whichever way you read it, this title reproduces a way of imagining the internet which emphasizes – indeed almost presupposes – the idea of *difference* between the uncharted virtual domains of cyberspace and all that is yet known. It presumes, but does not demonstrate, that the internet self-evidently affords a novel and separate territory, which has certain properties and possibilities all of its own. It suggests that the world in the internet age will unquestionably be new, the commentator's task being to outline whether it will be literally or ironically brave. It carries the danger of overlooking the connections and continuities between the life in the age of the internet and mundane, familiar, real life. This volume should not, however, be singled out for exceptional criticism. Much of the recent explosion of analytical commentary accompanying the growth of the digital information and communication technologies (ICTs) has focused on their putative potential to transform society radically. This has rendered the internet a key terrain for the expression of manifold concerns about the self, culture, community, economics and governance in late twentieth and early twenty-first centuries. Moreover, this commentary has by now solidified into a distinctive and well-established analytic milieu (into which the title *The Internet: Brave New World?* fits very well), with its own language, precursors, expectations and assumptions, which together establish a particular, largely dominant, mode of imagining what the internet is and what it offers. It is, I think, important to interrogate this mode of thought.

IMAGINING CYBERSPACE

While there is not enough space in this essay to undertake a full mapping of internet-related social commentary, nevertheless, there are a number of key ideas which can safely be said to have emerged in the literature. One is the idea of 'cyberspace' and the related notion of 'virtuality'. The word 'cyberspace' was coined by William Gibson in his science fantasy novel *Neuromancer* (1984). He used it to describe the 'consensual hallucination' of a computer network-borne virtual realm, which the story's cyberpunk protagonists chose to inhabit. However, nowadays this word has a much wider currency and is quite casually applied in many discussions about the internet, the world wide web, virtual reality, e-mail, bulletin boards, chatrooms, multi-user dungeons games and so on, to describe a discrete and distinctive domain, supposedly created by ICTs. Moreover, people are said to enter and inhabit this terrain simply through using these technologies. As Sherry Turkle puts it: 'For many of us, cyberspace is now part of the routines of everyday life. When we read our electronic mail or send postings to an electronic bulletin board or make an airline reservation over a computer network, we are in cyberspace' (*Life on the Screen: Identity in the Age of the Internet*, 1995).

But a significant slippage is going on here, where a term created for the purposes of a work of fiction has become a commonplace device used for discussing people's actual, non-fictional, technology- related activities. This is, of course, problematic, not least as there is no automatic reason to assume that Gibson's mid-1980s imagination has any real congruence with the lived experience of internet use in the late 1990s and early 2000s. But the legitimacy of using the term in this way is too frequently not addressed; a fact which has had a profound effect on the way discussion about the internet has tended to be framed. One clear example of this sort of thinking can be found in Bell's introduction to the *Cybercultures Reader*. Bell highlights the need 'to understand the ways in which cyberspace as a cultural phenomenon is currently being experienced and imagined' and accordingly acknowledges the need for commentary to allow for and explore broad and plural definitions of cyberspace. He then goes on to argue that in order to understand cyberspace, two histories need to be traced. One is of 'the impacts of science fiction on the ways cyberspace works for us', a history very much shaped by Hollywood visions of techno-futures and the other is of 'the experiential, subjective sense of the hallucinatory (wired) world we are engaging with' (*Cybercultures Reader: A User's Guide*, in D. Bell and B. M. Kennedy (eds), *The Cybercultures Reader*, 2000).

The problem here is that this approach problematizes the issue of the social and cultural effects of ICTs in certain specific ways, which allow some questions to be asked, while marginalizing others. It allows one to ask, for example, what and where cyberspace is, where we are when we are in it, who we are when we are in it, what cultural influences give it its shape and what different cyberspaces are created by different users and their different usages. But it does not allow for the possibility that for many people, internet users and non-users alike, the idea of 'cyberspace' is largely meaningless. Certainly, everybody's experiences and understandings of their engagement with ICTs will be mediated through received cultural scripts and representations, but these need not have anything to do with *Star Trek* or *The Matrix*. And while Bell highlights his own experience of typing at a computer as being peppered with 'cyberpunk moments, cybertheory moments, tech-*noir* moments [and] cyborg moments' there remains no reason to think, as the implication seems to be, that his experience is typical or at least familiar enough to be given as an example.

Moreover, it is important here to take account of how the cultural reference points one brings to a given activity will shape not only one's understanding of it, but also one's experience of it. Personally, when I use the internet, I have no sense of being removed into 'virtuality', no sense of being in 'cyberspace', and overall the experience I have is a fairly concrete one. But then I do not much care for science fiction or for American blockbuster movies in general, which in turn means that the toolkit of cultural representations they afford have little currency in my life experience. Consequently, the metaphors that I tend to employ for understanding and experiencing the internet are rooted not in Hollywood (cyber)futures, but in familiar communication technologies like the telephone, the television, the library and letter writing. Hence, I would suggest, the subjective experience of engaging with a 'hallucinatory (wired) world' is not an essential, universal aspect of using the internet and related technologies, but is an effect of using them after having grounded one's understanding of them within this sci-fi frame of reference.

VIRTUAL IDENTITIES?

It is important to make this distinction because it relates closely to another key idea emergent in the literature, which is that the internet and other 'virtual' technologies will lead to a world in which people have new understandings of their relations to themselves. It is commonly suggested in the literature that ICTs afford unprecedented opportunities for creating and recreating new identities and multiple identities, freed from the constraints of geography, race, gender and embodiment in general. One key text in which this theme is heavily featured is Sherry Turkle's *Life on the Screen*, exemplified in her account of Doug, a young man who plays in a number of MUDs (Multi-User Domains, also known as Multi User Dungeons; these are internet sites which people from all over the world log onto and in which they play fantasy games akin to Dungeons and Dragons). Doug in fact takes part in three different MUDs, and plays four different characters. One character is a seductive woman, one a macho male, one is a rabbit and one is some sort of non-specific animal whose primary characteristic is that it has sex with other animals. But, central to Doug's experience is the fact that he plays all these characters simultaneously, keeping all the different MUDs on his computer screen continuously, splitting his attention, and apparently his sense of self, from one window to another, engaging with the different realms as they demand it. And so for Doug, the real world (which he refers to as 'RL' – real life) is just another window, just another screen to be engaged with, but he does not see it as his primary, defining reality. Other commentators have suggested that for many people the boundaries between themselves and the technologies they use are becoming fuzzy, reflecting a computer culture in which the physical body is so much 'meat', sometimes called 'wetware', the unreliable organic counterpart to hardware and software, a corrupt and corrupting barrier to the true 'virtual' expression of the computer-mediated self (D. Lupton, 'The embodied computer user', *Body and Society*, 1 (3–4): 1995).

The vision here is of technological advance offering deliverance from a constrained and constraining physical reality and of the cyborg –

a synthesis of the human and the technological, where subjectivity is dispersed through circuitry, where selves are no longer fixed by the boundaries of the physical body, but are freed by electronic networks to formulate, reformulate, mutate and explore themselves, at will, ad infinitum.

Once again, this way of imagining ICTs is problematic. One obvious problem is that cyborg or not, RL (real life) must remain the primary window until a 'virtual' technology is invented which removes the need to eat, drink, sleep and go to the toilet. Another problem is that, however liberating the technology may be, online opportunities to reconstruct the self remain constrained by a variety of non-technical factors. Just as in real life, interactional environments on the internet tend to be regulated by social rules. But perhaps more than this, the problem with this vision is that it is not new. In the guise of a revelation of something radical, what we actually have is a restatement of some rather familiar ideas. Deborah Lupton, for example, suggests that the contrast found in cyborg visions between weak flesh and authentic computer mediated subjectivity, is a contemporary reassertion of post-Enlightenment ideas about the separation and relative status of mind and body. Similarly, Kevin Robins sees in Jaron Lanier's visions of virtual futures a mundane rearticulation of the Kantian transcendental imagination (K. Robins, *Into the Image: Culture and Politics in the Field of Vision*, 1996), and Michael Benedikt sees in the discourse of cyberspace a modern restatement of ancient imaginary and mythic spaces such as Shangri-la, in which the bounds of physicality and time are transcended (*Cyberspace: First Steps*, in M. Benedikt (ed.) *Cyberspace: First Steps*, 1992). The technology may be novel, but the imaginations brought to bear upon it reflect long-standing human concerns.

VIRTUAL COMMUNITIES?

This same backward-looking futurism is also revealed in the articulation of a third key theme within the commentary, virtual community. Like cyber-identities, virtual communities are imagined as being (at least potentially) free from the limits and errors that dog non-virtual communities. Commonalities of interest and vision will be the organizing principles of collective life online, displacing physical and political geography. Cyber-selves, free of pre-given real life social rules and conventions, will write new cyber rules of interaction. But, this notwithstanding, the way virtual community is commonly imagined has some less than radical underpinnings. Just as 'cyberspace' traces back to *Neuromancer*, so we can trace the phrase 'virtual community' back to one seminal text by Howard Rheingold, in which he gives an account of his involvement with an electronic community (the WELL – the Whole Earth 'Lectronic Link) based around the San Francisco Bay area in California (*The Virtual Community: Homesteading on the Electronic Frontier*, 1993). And the vision he gives is one in which, although the *means* of producing a community may be new, the values, principles and ideals of what a community could and should be are rooted in a nostalgic remembrance of real world (American) community, once possessed but now lost.

But not only are these ways of imagining ICTs and the internet old-fashioned in terms of the points of reference they employ, they are also old-fashioned in as much as the key ideas – cyberspace and virtuality, fluid and multiple computer-mediated identities and virtual communities – first came into common use in the early 1990s, a time when these technologies, and the internet in particular, were the province only of an interested and enthusiastic minority, which was simultaneously unified within itself and somewhat separated from the rest of the world by its members' shared relative technophilia. Hence, these key themes, of separateness, of newness, of unboundedness, did provide meaningful analytic metaphors to engage with the (quite specific) experiences of early users. But since the advent of the point and clickable world wide web, with a love and understanding of technology no longer a prerequisite of usership, the internet has moved inexorably towards becoming a mass medium. And as this shift has taken place, so an ever-increasing number of disparate imaginations and practices have been applied to make sense of and engage with this technology.

There is clearly a need, then, for commentary on the social effects of the internet and related technologies which eschews the 'cyber' school of thought. Recently, this need has been met in part by work undertaken under the Economic and Social Research Council's 'Virtual Society?' programme, which ran from 1997 to 2001. This programme was characterized by a commitment on the part of its members to pursue research with an attitude of what Programme Director Steve Woolgar called 'positive scepticism' (see, for example, the introduction to the 'Virtual Society?' programme's *Profile 2000* document, available at www.virtualsociety.org.uk). Positive scepticism begins by asking not *how* but *whether* the internet will change the world, not *how* but *whether* 'cyberspace' is experienced by disparate users of ICTs.

THE INTERNET IN THE REAL WORLD

The findings of both research pursued within the 'Virtual Society?' programme and other work undertaken with a similar, 'positively sceptical' attitude, give images of the 'internet age' which run counter to what one might have expected. One counter-image is that the internet, it seems, is not transforming society. It is influencing and changing it, but not in a truly radical way. There are a number of reasons to think this. For one thing, it seems that new technologies do not displace old ones, but rather are used in a supplementary way. E-mail, for example, has not swept all before it, but has simply joined memo writing, the telephone, letters and talking face to face as one communication technology among many (S. Woolgar, 'Virtual Society? Beyond the Hype?', *The Source Public Management Journal*, 2000). But another, perhaps more important, factor is that although the number of people accessing the internet is still rising, there is evidence suggesting that this increased uptake is cautious, rather than zealous ('Why users are falling out of love with electronic communication', *The Guardian*, 20 June 2001, see also BBC News Online, 'E-mail slips as web use grows', 19 June 2001). Moreover, it seems that some people are even rejecting the internet outright. A significant proportion of American adults, by no means confined to those who have never tried the technology, have been reported as seeing no use in the internet and having no intention of getting online ('Slump in number of US internet users', news report from *Cyber Dialogue*, 20 November 1999, available at www.nua.ie/surveys/).

Another counter-image is that the internet is frequently not virtual. Like the notion of 'cyberspace' just discussed, the idea of 'virtuality' confers a sense of internet life being somehow separate and discontinuous from ordinary life, a quality often seen as being somehow an inherent property of the technology. But empirical work examining how different groups of users engage with the internet suggests that more often than not, usage is very much grounded in the 'real' and not the 'virtual'. Daniel Miller and Don Slater's work, for example, strongly suggests that the internet is not virtual in Trinidad (*The Internet: An Ethnographic Approach*, 2000). Their anthropological study showed that despite an extensive and enthusiastic adoption of the internet among Trinidadians, their usage and understanding of it did not reflect any significant distinction between the 'real' and the 'virtual'. Rather, the potentials and applications of the new technology were assimilated and integrated into and made sense of by reference to pre-existing Trinidadian understanding and practices concerning work, family, friendship, leisure and national identity. Indeed, so complete was this assimilation, that the internet was not seen as something apart, but as something which had a close affinity with the complex of social meanings by which Trinidadian identities have been historically and habitually articulated. In other words, the internet was understood as 'naturally Trinidadian' and as an equally 'natural' medium for the global expression of certain factors central to Trinidadian identities – national pride, cosmopolitanism, entrepreneurialism, understandings of family long rooted in the need to communicate over large distances, local forms of casual chat and so on. Overall, then, the arrival of the internet in Trinidad, mediated as it was through a context which emphasized continuity rather than difference between the pre and post-internet ages, did not result in Trinidadians becoming disembedded from their existing senses of place, locality and identity, but actually did much to extend and reinforce them.

Neither is the internet necessarily virtual in Britain. My involvement with the 'Virtual Society?' programme was as a researcher on a project looking at sites which provided public access to the internet, in order to explore how they fostered social inclusion. We studied a selection of internet cafés (which are generally small businesses, located in urban areas, where people can get access to the internet for a modest fee) and 'telecottages' (usually voluntary sector or community organizations, generally rurally located, again offering access to computers for local people). And, similarly to Miller's work, what we found was that the way staff and users alike engaged with the computers and the internet was very much shaped by local context, very much grounded in senses of place, self and reality which were predicated on previously existing, familiar factors such as geography or non-virtual friendship, kinship and community networks. This was manifested in part by the way the sites were organized, as they tended to be more successful when they had effective links to local non-virtual social, business and community networks. The most successful cybercafé businesses tended to be those which exhibited the same properties as successful ordinary cafés; a pleasant social and physical environment in which to spend time, established in a good location (that is, a manifest physical link to the local business community). The most successful voluntary sites tended to have connections with other local community organizations, perhaps by inhabiting the same premises or being located in already existing community centres or by having staff with good inter-organizational personal contacts.

Moreover, the driving vision behind many 'telecottages', far from seeking to unfix and radicalize the identities of those who used them, was the contrary aim to encourage economic and social regeneration within a particular physical, geographically defined community. Indeed, often the primary virtual presence that such sites exhibited would be a website whose major concern was not to pursue the creation of a new technologically enabled 'virtual' community, but rather to use the internet as one tool among many to represent, re-establish and reaffirm existing ideas about their real world community.

And the same 'real world' groundedness was reproduced in the characteristics of the users of these sites. For one thing, users tended to live locally. Even in the urban internet cafés, most of the users were not, as one might have expected, Antipodean backpacking citizens of the world, the vast majority living within ten physical miles of the site. For another, the uses and perceived usefulness of the medium were reckoned in terms of the 'real world'. The overwhelmingly dominant use that was made of public internet access sites was one-to-one communication via e-mail, in which by and large it appeared that single, physical, fairly unitary selves, determined and fixed by many familiar non-virtual factors such as age, gender, ethnicity, place of birth and so on, would communicate with other similarly fixed selves within the confines of equally non-digitally mediated networks and concepts of 'real' friendship, kinship and community. The fewer number of users who did access internet sites tended to do so either to further some or other non-virtual interest – a football club, a favourite music group, a hobby, dates and times of local events – or to make online purchases. Few people in the sites we studied were interested or engaged in those online activities which might have allowed for the more radical possibilities highlighted by the 'cyber' mode of imagining the internet. Very few users wanted to generate their own content, very few used chatrooms. In sum, then, in the sites we studied the 'real' and the (supposedly) 'virtual' were on the whole not seen as discrete terrains, in as much as that 'virtual' technologies were usually only employed towards furthering some or other 'real' end. Overall, then, once again the picture is one in which these new technologies have not engendered something radically new, but have been incorporated into people's already existing understandings of self and community.

Neither is the internet necessarily virtual online. Jason Rutter undertook a six-month ethnography of an online newsgroup (RumCom.local) and his results support the contention that analysis which concentrates on internet spawned opportunities for identity play is highlighting an atypical form of usage, a vestige of the time when the internet was a novel and minority medium (Rutter *et al.*, www.virtualscoiety.sbs.ox.ac.uk/GRpapers/rutter). He found that once the internet-centred community he studied had become well established and online interaction within it was no longer a novelty, the participants abandoned the possibility of articulating fluid multiple selves and sought deliberately to fix themselves in the eyes of other participants. They did this in large part through explicitly integrating their online and off-line identities. Rutter identified various strategies that participants used in their interactions by which they achieved this fixing of self, such as proffering information about their non-internet lives (their history, their biography, their physical state, what they watched on television last night and so on) and by using 'signature practices' – unique appendages added to the end of the messages they sent, indicating their authors. Moreover, these practices were not simply a matter of preference. Significantly, Rutter contends that this identity fixing and real world contextualizing activity was essential for the establishment of trust between newsgroup members, an achievement without which serious, ongoing interaction would never be maintained or online community created.

Overall, then, once again Rutter's research suggests that to forge an understanding of the internet and related digital information and communication technologies based on the assumption that the 'real' and the 'virtual' are separate, discontinuous territories, each with their own distinctive social properties, each requiring their own discrete forms of self, would misrepresent the reality of online life for many – and perhaps most – people who use these technologies.

Furthermore, all this research counsels against the technological determinism implicit in many accounts of life in the age of the internet; the idea that new technical possibilities will lead to a new world, new community, new humanity. But what we see from these studies is the nature of the internet, what it means, how it is used and so on, being shaped locally by different groups of people, according to their disparate needs, beliefs, desires and worldviews. So the lesson is that if we want to know what our wired world will look like, we must look beyond the technology, towards the factors that shape the people that shape the internet.

This is where my subtitle comes in. Keen-eyed readers will have spotted that I have made a reference here to Stanley Kubrick's 1964 film *Dr Strangelove*, a tale of cold war paranoia. I think I need to explain the connection, which is, if I am honest, tenuous – but probably no more so than connecting the internet to Huxley's *Brave New World*. (And after all, it is only a title.) Anyway, the point I want to make is that there are some parallels between the internet and the harnessing of the power of the atom. Both technologies seemingly defined the spirit of their respective ages. Both technologies also defined the hopes and anxieties of their respective ages. And both technologies reflect, and have served to extend and reinforce, the politico-economic configurations of their respective ages.

This point is drawn out clearly in Manuel Castells's *Network Society* trilogy (*The Rise of the Network Society*, 1996; *The Power of Identity*, 1997; *End of Millennium*, 1998). His argument, briefly, is that while the internet now underpins the distribution and processing of information in all realms of society, it is the necessary technical basis for the 'new economy', a new form of capitalism, now developing globally. Castells is not simply talking about the dotcom companies here, but about the increasing centrality of the internet and networking to all kinds of business and economic activity. And it is the development of this 'new economy', he suggests, not the supposed existential possibilities of selfhood associated with ICTs, that has spawned the current interest in the internet shown by governments, companies and society in general. The internet age, then, is the age of global consumer capitalism. Hence my subtitle, like the title of this book, can then be read either literally or ironically. Given the centrality of ICTs to Castells's new economy, whether or not we should stop worrying and love the net depends on how worrying we find global capitalism. Castells himself outlines a number of challenges and difficulties that the new internet-mediated economy will bring: 'systematic volatility' in financial markets, increased social inequality and social exclusion and eventually the need to reconstruct the instruments and institutions of state, to name but three.

CONCLUSION

I would like to return to the question of identity, so central to the 'cyber' imagination, because I suspect selfhood in the internet age will reflect more than anything else the dynamics of capitalism. Writers such as Rose have highlighted how governance, identity and capitalism are all linked through the contemporary emphasis on self-realization through consumption (*Governing the Soul: The Shaping of the Private Self*, 1989). His argument is that, nowadays, people understand themselves primarily at an economic level, as individual consumers, constantly incited to render their lives meaningful through decisions of personal consumption. In such a world, individuality becomes a function of buying (either literally or metaphorically in the sense of 'buying into') an off-the-peg, mix and match lifestyle. Moreover, political citizenship and social inclusion become functions of the free exercise of personal choice within the marketplace. The world wide web is already starting to reproduce the conditions which will propagate this sort of selfhood. People bring their understandings of themselves as consumers of lifestyles to the internet and respond most favourably when the internet responds in kind. It is, for example, not insignificant that the *Which?* Online Annual Internet Survey 1998 found that widespread and various worries about the internet were often assuaged once people actually got online and realized the benefits the internet offered them *as consumers.*

Similarly, Wilbur has argued that the commonality of interest which draws internet communities together may often best be understood as a commonality of consumption preferences (S. P. Wilbur, 'An archaeology of cyberspaces: Virtuality, community, identity', in D. Porter (ed.), *Internet Culture*, 1997). To be sure, people do not just use the internet to make purchases. But in a world where more and more the discourse of the individual consumer shapes how people know themselves, that understanding will more likely than not underlie how they engage with and understand the technology, when they seek and create content and when they communicate and commune with each other. The 'brave new world' of the internet will have more than a passing similarity to the tired old world from which 'cyberspace' was supposed to deliver us.

DEBATING MATTERS

Essay Two

COPYRIGHT AND THE NET
Chris Evans

The internet has made it incredibly easy to distribute information. Some people have welcomed this for the potential it offers for freedom of speech. Others have condemned it for the freedom it gives to pornographers and racists. But perhaps the most unexpected battle about internet freedom is that over copyright. The ease with which images and especially music can be copied and distributed over the net has led to a heated conflict. On one side are those who seek to preserve (or even extend) traditional intellectual property relations. On the other side are those who seek to exploit the new medium to its fullest extent.

In particular, the music compression technology of MP3 has met some powerful condemnation. Attempts which have achieved varying degrees of success have been made to outlaw it; to block hardware that uses it; to shut down websites that use it; and to legislate against translating other formats into it. The conflict centres on the role which intellectual property plays in the information age. In practice it takes the form of a clash between public and private interest. It has consequences that threaten to stifle the development of technology as well as curtail some of the liberties that we have traditionally enjoyed in the off-line world.

INTELLECTUAL AND MATERIAL PROPERTY

Intellectual property, like all forms of commodities, is the product of labour. Musicians, authors, film makers, and programmers can work for months or even years to produce music, books, films and software. The fact that this labour is intellectual rather than physical does not alter the fact that it initially involves time and effort and should be financially rewarded. However, the intellectual origin is often obscured by the fact that the distribution of the product takes a material form. When you buy a music compact disk (CD), for example, you do not usually buy just the music. You also buy the plastic and aluminium disk in which it is embodied.

At first glance, intellectual property appears to operate in similar ways to material property. Property law places ownership of a material commodity with an individual. It is against the law to take possession of someone else's car, for instance, without their consent. Likewise it is against the law to take possession of someone else's intellectual property, a song or a piece of software for instance, without their consent. If you delve a little deeper, however, you will discover that material and intellectual property are very different and that the laws that define them operate in very different ways.

Take material property first. While it is a crime to take possession of someone else's material property without agreement, it is not against the law to own an identical commodity. So there is nothing to stop your neighbour owning a silver Toyanda Civica car that is identical to yours. Likewise there is nothing to stop Toyanda manufacturing as many identical silver Civicas as it likes. Most importantly, putting intellectual property issues aside, there is also nothing to stop someone else manufacturing a similar car to the same specification and selling it at a lower price. That is the basis of competition in the market. And finally, if you decide to sell your material commodity to someone else, you cease to be the owner and the commodity becomes theirs.

The story is very different for intellectual property. It is a crime to take possession of someone's intellectual property without agreement. However, it is also a crime to own an identical commodity without the owner's consent. Copyright law prohibits me from possessing a copy of one of a recording company's songs, for instance, without their consent. They can sell as many copies of their songs as they like. However, it is a crime to make or attempt to sell an identical copy (a practice that goes by the name of 'piracy'). Finally, you can sell one of your songs to someone, but normally you continue to be the owner and the recipient typically cannot reproduce the song and sell it to anyone else. The selling arrangement is thus more akin to a licensing agreement. The only way to give up ownership of intellectual property completely is to surrender not just the song itself but also the *rights* to it.

When you buy a music CD, you are thus buying both the material from which it is manufactured and the (licence to play) songs it contains. The difference between material and intellectual property means that you end up owning the former, but the music company continues to own the latter

ECONOMICS OF COPYRIGHT

One of the distinguishing characteristics of intellectual property, especially with the developments of modern technology, is that it is very easy and cheap to reproduce it. Books and articles are easily photocopied or faxed. Music on CDs is easily extracted and sent across the internet. Television programmes are easily videotaped. Even with recent developments in technology, the same cannot be said for most material property. While the cost of manufacturing motor cars continues to fall in real terms, it still takes time and therefore expense to reproduce them. Most people could not begin to imagine what it would take for them to produce a copy of a leading brand of car, for instance, but it would not take more than a few minutes for the computer literate to produce a copy of a leading brand of software.

In fact it is the material embodiment of intellectual property that makes most of the profit. Rather than the artists, it is the labour of bauxite miners, Asian oil workers and English packers that lines the pockets of the CD industry. For example, Polygram Records outsourced the packaging of its CDs to M&S packaging in 1996. In their warehouse in Blackburn, 180 workers packed a quarter of a million CDs every day. Shifts were 12 hours long and casual workers were paid just £3 an hour. It is under these exploitative conditions that record companies are able to maintain their profits. From the consumer's perspective it is not the plastic and aluminium that matter, but the music the CD itself contains. From an economic perspective, the reverse is true. The music is more the medium by which millions of bits of plastic and metal are shifted into the hands of music fans the world over. This economic difference lies at the root of the social and legal treatment of intellectual property. In order to understand how the situation has arisen, however, it is useful to look at how copyright evolved.

THE HISTORY OF COPYRIGHT

Copyright is a comparatively modern invention. There was no copyright law before the development of the printing press. People were free to copy stories usually by memorization and word of mouth and occasionally through painstaking handwritten duplication of manuscripts. No laws prohibited this, although owners often controlled copying by controlling access to the manuscripts (as did the monks described in Umberto Eco's *The Name of the Rose*). Copying was a difficult, time-consuming and thus expensive activity.

With the development of the printing press by Johannes Gutenberg in the 1450s, the reproduction of printed works became quick and cheap for the first time. Church and government initially attempted to control the type of content that was reproduced by forming pacts with printers, granting them monopolies in return for pledges not to print seditious or heretical works. Printers were licensed by the King through an organization known as the Stationer's Company. Printers would enter the names of books they intended to publish in the Stationer's Register and company rules prohibited anyone else from printing them. There was no real copyright law, however. There was nothing to prevent printers publishing any material they could lay their hands on provided it was not already entered in the Stationer's Register. They also did not see hand copying as an economic threat, as it was much more time consuming (and therefore expensive) than printing, so it was not prevented.

There were two problems inherent in this company arrangement. On the one hand, as monopolies, the printers were insulated from the full force of the market. They could thus set the prices of books artificially high without fear of being undercut by competitors. On the other hand, the reproduction of printed material was so cheap that they could pay authors next to nothing for it. As long as there was a ready supply of written material there was no economic reason to pay anyone for it, much as many internet users find little economic reason to pay much more than the cost of a blank CD for digital music today. Printers generally only paid authors honoraria as an inducement to give up their material. There was therefore little financial incentive for authors to write new works, as the compensation was too low. The market system failed to provide the conditions for the reproduction of intellectual property and the monopolies could not help.

With the Enlightenment in the eighteenth century, the situation was exacerbated. The demand for published works accelerated and the control of the monarchy was weakened. Authors began to object to getting little or no compensation for their work. People began to object to the high prices of books fixed by the printing monopolies. Would-be publishers objected to being blocked by government licensing. Freedom of speech and the demand for universal access to information challenged the company licensing arrangements.

In order to address these problems parliament passed the first copyright law in 1710, known as the Statute of Anne. The aim of the statute was to address the economic problems of the reproduction of books. In particular the fact that authors got paid little yet prices were very high:

> Printers, booksellers, and other persons have of late frequently taken the liberty of printing, reprinting, and publishing, or causing to be printed, reprinted, and published, books and other writings, without the consent of the authors or proprietors

of such books and writings, to their very great detriment, and too often to the ruin of them and their families.

(Statute of Anne, 1.8.8, 1710)

The statute made it very clear that it was intended 'for preventing therefore such practices for the future, and for the encouragement of learned men to compose and write useful books.'

The statute, therefore, had to balance three things. It had to provide for compensation to authors in order to encourage them to write books in the first place. It also had to somehow weaken the printing monopoly in order to force the prices of books down. And it had to do both these things while ensuring that it was commercially viable to print books at all.

The mechanism used to do this was twofold. First, the statute granted rights to authors rather than printers. Authors could take legal action against anyone who copied their work without their consent. This gave writers the power to demand greater compensation for their works. Second, it limited the timescale under which those rights applied. Previously printers maintained a monopoly indefinitely which allowed them to charge high prices. Under the statute, the copying rights of the author lapsed after 14 years, after which anyone could reproduce the work. This was intended to have the effect of forcing the price of copyrighted books down to bring them more into line with the prices of books for which the copyright had lapsed (and which were subject to the full force of market competition). It also had the effect of creating a temporary monopoly for printers in association with an author, which made the printing of books a potentially profitable activity.

THE TREND TO EXTEND COPYRIGHT

Since the Statute of Anne, the length of copyright protection has steadily increased. In the USA it has been lengthened 11 times in the last 40 years. It had been lengthened, by 1997, to the life of the author plus 50 years. It had also been extended to a much broader range of creative works including paintings, sculptures, photographs, plays, musical works, sound recordings, television programmes and films. In the UK the most significant recent copyright legislation is the Copyright, Design and Patents Act (1988). Like in the USA, this set the period of copyright to be the life of the author plus 50 years.

In 1998 the US Sonny Bono Copyright Term Extension Act lengthened the period yet again to life plus 70 years. Company-owned copyright was extended from 75 to 95 years, as were all copyrights for works produced before 1978. The act was spearheaded by Disney to protect the rights of Mickey Mouse, Pluto and Goofy, which were scheduled to enter the public domain in 2004. Few could see the public benefit of this act, however. For example, David Post, a professor of law at Temple University, commented: 'It's a disgrace. There is no better example that I can imagine, literally, of Congress caving in to small, highly focussed special interests. There is no conceivable public benefit from the additional 20 years. Zero' (*Reason*, March 2000).

Since the Statute of Anne, then, the legal power of the entertainment and software industries to claim copyright on material has grown greater. Conversely, the legal ability of consumers to disseminate intellectual material has decreased. The Statute of Anne was aimed at 'the encouragement of learned men to compose and write useful books' and sought to reduce the monopoly of printers. By contrast, recent extensions have been aimed at protecting the private interests of the industry and extending their monopoly.

INTRODUCTION OF INTERNET TECHNOLOGY

The development of the internet has further increased the tensions between the private interests of publishers and the general public interest that have existed since the Statute of Anne. It is now possible for any individual to publish material as a web page for little more than the price of a telephone call. Moreover the size of the potential audience for such material – web users around the globe – is greater than ever before. With modern compression techniques it has also become possible to distribute music files and digital videos in a matter of minutes. Consumers have gained unprecedented access to all kinds of materials. The cost of distributing a book or even a video has become tiny. As a consequence the monopoly prices previously charged by publishers, and now the entertainment and software industries, appear inflated. Not only that, but the role of the industry as middlemen between authors or artists and consumers is open to challenge. The web creates the possibility of direct trading. Or at least creates opportunities for dotcoms to set themselves up as publishers and producers in direct challenge to those industries that have so long relied on paper-based books and plastic disks for their revenue. Once more, consumers want cheaper prices, competitors want to challenge the monopolies and the old industries desperately seek to maintain the status quo.

THE RESPONSE OF THE RECORDING INDUSTRY

The response of the recording industry to the impact of the internet has been twofold. First, they have been working hard on digital technologies that make it difficult if not impossible for users to copy files. Second, they have pursued every legal means to clamp down on what they see as the violation of intellectual property rights through prosecutions and the introduction of new legislation.

Digital music distribution represents one of the best examples of the clash between technology and copyright. The development of the sound compression technology, MP3, for instance, was met with attempts to outlaw it as it made it easy to distribute music over the net. Rio Systems who manufactured the first MP3 player, for instance, were victims of an unsuccessful lawsuit brought by the music industry that accused them of facilitating copyright violation.

THE CASE OF NAPSTER

Probably the most famous of battles over the impact of the internet on intellectual property is that over Napster. Napster began life as a very simple piece of file-sharing software. The idea of Napster was for internet users to share digital music files over the internet. They simply encode music in the high-compression format MP3 and place their files in a particular folder on their hard disk. They then connect to the internet and start up the Napster software on their machines. The software collects details of their music files and adds it into a central database stored on Napster's own server. Napster users everywhere in the world then have access to music files on everyone else's hard disk. They can search Napster's database for any particular file they want. When they have found it, the database simply tells the user where to find it on someone else's hard disk. The software then sets up a direct internet connection to the disk, this time without going through Napster's own server. They can then download the file in a matter of a few minutes or even seconds. As Chief Executive Officer (CEO) Eileen Richardson put it: 'Everyone looks at Napster and goes, "Holy Shit!"' (*Practical Internet*, issue 39).

The actual data are not stored on Napster's server so the company has not actually copied the material themselves. However, there is nothing inherent in the technology to prevent any user from placing copyright material in their shared folder. And that, of course, is what many users did. Having purchased a music CD from their favourite band, they would encode it in MP3 format (a process known as 'ripping') and put the files in their shared folder. Fellow Napster users would benefit from being able to download the music without paying the price of the CD and they themselves could do likewise with other music files.

From the user's point of view they are simply sharing their CDs with their friends. Except, of course, that in this case their friends turn out be every other Napster user on the globe. In any case, in the face of the technology of the internet, CDs appear to them to be prohibitively expensive. If it costs little more than the price of a blank CD to make your own copy, why should you be expected to pay 10 to 20 times that amount to buy it from a shop? From the music industry's viewpoint, users are illegally copying copyright material and are forcing them potentially to lose out on sales of CDs. And Napster appeared to be helping them.

This last view was precisely that taken by the Recording Industry Association of America (RIAA), which sued Napster for facilitating breach of copyright. Initially, Napster argued that it was not responsible for copyright violation since they did not store the files and could not be held responsible for whether users have legal or illegal copies of music. It was like, they argued, suing pencil manufacturers because pencils can be used illegally to copy copyright materials. And in any case, they argued, overall CD sales had actually increased. The RIAA countered by claiming that CD sales had not increased as much as they might have done were it not for the likes of Napster and that there was a substantial trade in illegal files facilitated by the use of Napster.

In the end, Napster lost out. The judge ordered them to introduce blocking software to cut down on the traffic of illegal files. By the time of the ruling in March 2001, Napster had already signed a deal with Bertelsmann (one of the original prosecutors) to set up a subscription service under which users had to pay a monthly fee. But file sharing did not stop with Napster. Quick to step into its boots were many alternative systems such as Gnutella, Aimster, OpenNap and FreeNet. Most of these lack the centralized server approach taken by Napster. Moreover, the source code for the systems is often freely available on the net. Taken together this makes it difficult for opponents such as the RIAA to target any individual company as being responsible for facilitating copyright violation.

THE CASE OF BEAM-IT

The digital music company MP3.com introduced a similar technology called 'Beam-It'. The concept is just as simple as Napster, but even cleverer. Launched early in 2000 it allowed net users to access MP3 music files from any computer, anywhere, anytime. As MP3.com CEO Michael Robertson described: 'This is so cool, it's going to blow you away' (*Practical Internet*, issue 39). The way it works is that net users would put a CD in their drives and details of the disk would be transmitted over the net to MP3.com. The company would then transfer MP3 files corresponding to all the tracks into a private folder belonging to the user on their server. With a password the user can then download and play the files from anywhere on the net. Since the user put the CD in, MP3.com are not technically giving them anything they didn't already have. Except the RIAA disagreed, since it might not have been their own CD or even their own password. MP3.com found themselves on the losing end of a lawsuit from the RIAA in April 2000. In addition to other legal cases like these, the recording industry has been developing ways in which technology can be used to enforce copyright.

DIGITAL RIGHTS MANAGEMENT (DRM) TECHNOLOGY

The major technical response to the impact of the information and communications technologies has been the development of digital rights management (DRM) technology. The idea of DRM is to provide technological control of how any given piece of intellectual property is used. Unlike in the past where once you bought a commodity it was yours to do with as you wished, the aim of DRM is to control exactly what you can do with it. This might include how many times you can copy it, when you can copy it, what you can play it on, how many times you can use it or even when it will expire and become unusable. Thus DRM aims to control the use of intellectual property in areas that legislation has never reached. As Lawrence Lessig wrote in *Code and Other Laws of Cyberspace*: 'Code can, and increasingly will, displace law as the primary defense of intellectual property in cyberspace' (2000). A good example of this is Rosetta Books, which plans to offer the Agatha Christie mystery *And Then There Were None* in a time-based permit edition. It will allow ten hours of reading time after which the content will no longer be available.

After the ruling in favour of the RIAA, Napster provides another example. Users who continued to use the service found that they not only had to pay a monthly fee, but also that in the future the MP3 files that they download will be converted into uncopyable NAP files.

RESISTANCE TO DIGITAL RIGHTS MANAGEMENT

The introduction of DRM has been met with opposition from many consumers. Thomas Claburn predicted it would have devastating consequences for the future of the net. In his work, *The End of the Web as You Know It* (June 2000), he lamented: 'See a Web article you like? Cut and paste, and presto – it's a part of your newsletter. Great-looking picture on that site? Right-click and it's your wallpaper. The wacky video file that makes you split your sides? E-mail it to 432 of your closest friends. Too bad those days are over.'

Marc Zeedar writing in *Copyright Wars and Digital Dreams* (February 2001) was equally critical: 'What copy protection schemes do is hurt legitimate buyers. I buy the e-book but can't read it or transfer it to another computer. I pay for a DVD but it won't work with my Linux computer. I spend money for a CD, but can't convert it to MP3 format to listen on my portable player.'

DRM means that net users in general will no longer have the ability to control the use of music in their own possession. In the off-line world we have been free to record the latest episode of *Friends* or tape a CD for personal use on the grounds of 'fair use'. But with the internet, DRM technology may prevent this. As Robin Gross of the Electronic Frontier Foundation commented: 'We are going to find that the kinds of liberties we've traditionally enjoyed in the real world are not going to exist on the internet.'

THE DIGITAL MILLENNIUM COPYRIGHT ACT (DMCA)

The success or failure of DRM may be decisive in the battle over copyright. However, the industry has been careful to observe that it has been technology that appears to have created their problems in the first place. It has therefore been active in finding every legal mechanism it can to back up the development of DRM.

Legal support for DRM to protect copyright comes in the form of the Digital Millennium Copyright Act 1998 (DMCA). The Act was designed to implement the treaties signed in December 1996 at the World Intellectual Property Organization (WIPO) Geneva conference. It attempted to update copyright legislation to take account of the digital age and is likely to have a significant impact on future copyright legislation throughout Europe. It was supported by the software and entertainment industries and opposed by scientists, librarians, and academics.

The most significant highlights of the act are that it:

- makes it a crime to circumvent DRM.
- outlaws the manufacture, sale or distribution of code-cracking devices used to circumvent DRM
- requires that 'webcasters', like radio stations, pay licensing fees to record companies.

The Act made it clear that the US Senate supported DRM technologies for the protection of copyright. It also put its weight squarely behind the act by earmarking $4 million for copyright prosecutions in 2001 and $10 million for 2002.

THE CASE OF DMITRY SKLYAROV

The first major victim of the DMCA was Russian programmer Dmitry Sklyarov who was arrested by the FBI in August 2001. Adobe Systems Inc. alleged that he was responsible for aiding the circumvention of their DRM system. His arrest followed a presentation he gave to the DEFCON 9 conference in the USA on the shortcomings of various encryption systems. Following protests in 25 major cities around the globe, Adobe Systems Inc. removed its backing for his prosecution, but the FBI pursued it. Sklyarov and his employer ElcomSoft were indicted on five counts under the provisions of the DMCA: four of trafficking and one of conspiracy to traffic circumvention technologies.

The apparent crime committed by Sklyarov and his employer was to develop software to allow Adobe's eBooks to be converted from one format to another. The software, called the eBook Processor, is perfectly legal in Russia where it was created. There is no suggestion that either he or his firm violated anybody's copyright. However, the DMCA makes it a crime not just to violate copyright, but also to develop technology that might be used by someone to violate copyright. The act makes this a crime regardless of the use to which it is put. As journalist Dave Amis put it: 'The eBook Processor can be used as an illegal circumvention device. Crowbars can be used for burglary. Kitchen knives can be used as offensive weapons. [But] there are no calls to criminalize crowbar or cutlery manufacturers' (Internet Freedom website, August 2001).

DIGITAL VERSATILE DISCS AND
THE CONTENT SCRAMBLING SYSTEM

The difficulty that industry has in controlling the use of digital copyright material and the lengths to which they will go to attempt to control it are well illustrated with the case of Digital Versatile Discs (DVDs). DVDs employ a DRM standard called the content scrambling system (CSS). Consumers need a key built into their DVD players or computers to view the contents of a disc. This allows the industry to control when and where a film can be viewed, and on what type of player.

The industry claims that this control is necessary to prevent loss of revenue by international sales of DVDs. For example, suppose a film is released in the USA in February. It comes out on DVD in the US in May, but will not be released in Europe and Australia in cinemas until June or in South America or Asia until July. If the DVD released in the USA can be viewed on DVD players anywhere, the movie industry say it will lose revenue from people who buy the DVD rather than going to the cinema. So CSS is used to prevent USA DVDs (known as 'region one') from working on DVD players in Europe ('region two') or elsewhere (regions 'three', four' and 'five').

On computers, the DVD Copy Control Association (DVD CCA) signed licensing deals with the two major manufacturers of operating systems, Microsoft and Apple, to allow DVDs to be playable on Windows and Mac platforms. CSS was then used to prevent DVDs from being playable on other platforms that had not reached any agreement.

Unfortunately for DVD CCS, the sale of multi-region DVD players, which could play DVDs from any region, soon outstripped single-region players. Software patches for Windows and MacOS appeared on the net to overcome region coding on those computers. Users of the open-source Linux operating system began investigating ways in which CSS could be overcome to allow them to play DVDs on their machines. By the autumn of 1999, 16-year-old Jon Johansen and his two associates had worked out how CSS encryption worked. They developed code that would allow the decryption of CSS files (DeCSS), which would help Linux users to play DVDs. In the spirit of open source software, they posted their code on the net. The movie industry and the DVD CAA were thrown into a panic. Their response was to resort to a number of legal actions in a vain attempt to remove the DeCSS code from the public domain. Of course, the global nature of the internet made this pretty much impossible.

The threat of DVDs being pirated was used as justification for these actions. But there was no evidence that DeCSS had led to piracy. As Don Marti of the Silicon Valley Users Group commented:

> The idea that this DVD CSS case is about illegally copying movies is completely false. If it had anything to do with illegally copying movies, we would see an illegally copied movie in evidence. So far, there isn't one bootleg DVD, there isn't an offer to sell or trade a bootleg DVD. There isn't so much as a chat room log about bootleg DVDs.
>
> (*Wired*, January 2000)

Norwegian police raided Jon Johansen's house to collect evidence stemming from allegations that he violated trade secrets to create DeCSS. Johansen and his father were also charged with publicizing the code on his father's company's website. They both faced fines and up to two years in prison.

In December 2001, the DVD CCA initiated a lawsuit in California against 72 individuals, accusing them of stealing trade secrets. In two federal cases filed in New York and Connecticut, the Motion Picture Association of America sued four individuals for contravention of the Digital Millennium Copyright Act.

 **CONCLUSION**

The original copyright legislation embodied in the Statute of Anne was clearly aimed at encouraging the dissemination of intellectual work for the benefit of all. The development of the internet has meant that the potential audience of any given piece of work is greater than ever before and that the cost of disseminating such information is less than ever before. You would think that such an enormous social benefit would be welcomed. However, the balance of legislation has shifted to protecting the private interests of the recording, software and entertainment industries at the expense of public interest. As Jesse Walker, associate editor of *Reason* puts it: 'Rather than promoting enterprise and speech, copyrights and trademarks often restrain them, turning intellectual property law into "protectionism of the culture industry"' (March 2000).

The real potential of the internet and modern technology is being stifled. Not only is information dissemination curtailed, but also the development of promising new technologies, such as MP3, are being threatened. As the DMCA and the case of Sklyarov showed, there is a dangerous risk avoidance trend that outlaws technologies that could be misused, regardless of the social benefits they might have. Enormous amounts of effort are being put into developing digital rights management technologies that are explicitly designed to make it more difficult for ordinary people to disseminate information.

The real measure of the benefits of any form of copyright control should not be 'Will it compensate the artist?' or 'Will it be profitable for industry?', but 'Is it in the general public interest'? It is time for a cold, hard review of just what we want copyright legislation to achieve. It seems certain that without a change in approach the current battles over copyright will continue. Law will become more restrictive, token prosecutions will increase, piracy will continue, liberties will be further curtailed and technological development will be stifled.

The technology of the internet needs to be embraced like its predecessor the printing press. We should not allow our lack of imagination, or a sense of history, to prevent us from unleashing its true potential. Instead, we should be looking to develop new business models and legislative approaches that allow the unfettered development of the internet in ways we may never have thought possible.

Essay Three

THE INTERNET: A MENACE TO SOCIETY?
Ruth Dixon

The internet truly offers a 'brave new world' in the most positive sense, in that it is the most powerful communication tool in history and can be used for business, education and entertainment. Far from representing a menace to society, the internet has the potential to enhance our knowledge and our relationships in ways previously undreamed of.

Above all, it offers every user the unprecedented opportunity to be a publisher. This has never been true of any other medium and represents a huge creative opportunity which should be both encouraged and exploited. Unlike traditional 'vanity publishing' or self-publishing, which both incur significant costs for the writer, the internet enables people to express their views at minimal cost, potentially to a worldwide audience of millions and provides the tools to enhance the likelihood of that content being accessed.

There are undoubted benefits in being able to communicate directly with people from around the world. Instant and real time access to people of all ages and backgrounds means that common interests can be discussed, horizons can be broadened, and tolerance increased between both individuals and communities. Mutual support systems can be developed for those who may be vulnerable and lacking off-line support. Children and adults alike can enjoy the opportunity to interact on a level playing field, regardless of many of the social, cultural, religious, geographical or potentially discriminatory obstacles that may inhibit them off-line.

A powerful illustration of the potential of the net can be seen in the story of Hero Joy Nightingale, who suffers from 'locked-in syndrome'. Now in her mid-teens, Hero is able to communicate only through a system of hand gestures which are interpreted by her mother. Nevertheless, she produces an online magazine, *From the Window* (www.atschool.eduweb.co.uk/hojoy/), which has included articles by Stephen Hawking, Kofi Annan and Helen Sharman, to name but a few. The internet has given Hero the chance to communicate on a level playing field – as Catherine O'Brien wrote in the *Daily Telegraph*: 'It is easy to see why the Worldwide Web has been such a liberating medium for Hero. Conversation with her, while fascinating, is also a slow and wearisome process. But when she communicates via email, her disability is invisible and her prose fluent' (2 February 1999).

Hero gives her own perspective on her website:

> *From the Window* is now one year old. It has restored my energy and optimism where clinical psychologists couldn't, taught me huge amounts about journalism and IT, connected me to many who also enjoy considered prose ... I have passion not passivity, demand a right to participate, to grow into a contributing member of our global society, to be an artist.

The potential of the internet to encourage people to grow into contributing members of a global society was acknowledged by three US federal judges in their ruling on the Communications Decency Act. In granting the Citizens Internet Empowerment Coalition's (CIEC) request for a preliminary injunction against the CDA, they stated that: 'It is no exaggeration to conclude that the Internet has achieved, and continues to achieve, the most participatory marketplace of mass speech that this country – and indeed the world – has yet seen.'

It is not only on the personal level that the internet can be a liberating medium. Stories abound of the importance of the net to those who live under oppressive regimes and have previously had little or no opportunity to communicate their views. For example, Internet Relay Chat was used to communicate internationally during the Gulf War in 1992 and during the coup against Boris Yeltsin in September 1993, where IRC users from Moscow were giving live reports about the unstable situation there. John Perry Barlow, in his essay 'A declaration of the independence of cyberspace' (www.eff.org/~barlow/Declaration-Final), affirms the same empowering effect of the growth of the internet:

> We are creating a world that all may enter without privilege or prejudice accorded by race, economic power, military force, or station of birth. We are creating a world where anyone, anywhere may express his or her beliefs, no matter how singular, without fear of being coerced into silence or conformity.

IDENTIFYING AND TACKLING THE PROBLEMS

It will be evident from these examples that the public and private benefits of this socially inclusive and interactive communication tool should not be underestimated. However, as in society generally, there is a small minority of people who will exploit the possibilities of an essentially neutral medium for criminal purposes and it is important both for the rule of law and to foster confidence in the medium by the majority of law-abiding users that such abuses are seen to be effectively tackled. Otherwise the internet will come to be viewed as some kind of anarchic Wild West and many people will be put off using it or allowing their children to use it. This in turn will deprive them of the many benefits of the internet for education, entertainment, business and communication and for the younger generation in particular could well leave them disadvantaged in school and in the workplace. Free speech and access to it will in fact be restricted, not protected.

As well as the individual consequences of being denied access to the internet, this in turn is likely to accentuate the more general digital divide between those with internet skills and those without. The issue of provision is a difficult one, especially concerning the hardware needed for internet access. There is a danger of the materially poor also being information poor. (At the same time we should not assume that everyone will want a computer and modem at home just because they can afford to have them.)

There is a clear need for provision of public internet access points and this should be encouraged, but 'public' needs to be defined more broadly than just meaning libraries. The statistics on book ownership highlight the fact that there is a considerable sector of society for whom regular library use is unknown or may even be intimidating. Therefore, it is necessary to consider where public provision can be located in order to reach those who are already information poor. Perhaps internet terminals could be placed in pubs, fast food outlets or even recreational areas in the workplace, although this last suggestion is clearly contentious and would need careful planning and supervision. (There may be some need for legal protection for the owners of premises offering public provision in the event that users access illegal or disturbing material.)

The convergence of the internet with other media, primarily television, will facilitate access on familiar and trusted platforms. This enables internet provision in a form that is less threatening to many potential users. But it does raise difficulties in the area of regulation, with the attendant dangers of the material either falling between two regulatory stools and not being dealt with at all or being dealt with in an excessive and unacceptable way. There will be a challenge, too, in dealing with consumers' expectations of their media experience – in the UK at least, television viewers are familiar with the concept of the nine o'clock watershed, and have a degree of confidence that certain types of adult material will not be accessible until that time. Once they are using their television set to access internet content, there may be a blurring of boundaries and a jarring of expectations which industry and government alike will have a responsibility to manage.

The risk of creating or exacerbating the digital divide exists on the international as well as the individual level. We talk casually about the internet being a global medium, while the reality is that in many developing countries the aspiration for 'universal access' is more likely to mean a telephone in every village than a modem on every desk. By way of illustration, an advertising poster recently on display on the London Underground states that '80% of the planet have never heard a dialling tone'. Given the potential for real time international communication that the internet offers, this represents a challenge for governments and for the global telecommunications industry.

On a more positive note, one can be optimistic about the possibility of the internet promoting increased understanding of, and in turn tolerance for other nations and cultures. However, in order for the internet to offer as wide an expression as possible for the ideas and identities of different individuals from a variety of countries and backgrounds, we need to resist the drive towards the internet being a predominantly or even exclusively English-language medium. If this were to happen it would impose severe limitations on both access and content. Even within an English-speaking country such as the UK it cannot be assumed that everyone is fluent in English. The internet community should therefore seek to counteract this trend both with content in other languages and with online real time translation programs, provided free to the end user where possible, for example through incorporation into the browser software. Facilitating communication in this way would also benefit the development of the internet as a global marketplace, a major objective of the UK Government and the European Union, providing that other requirements are met to foster consumer confidence, such as adequate security measures.

In the context of enhancing the individual, societal and global benefits of the internet it is essential to have reasonable, workable and transparent mechanisms for tackling criminal content in the first instance and for empowering users to deal with material which they consider to be offensive or potentially harmful.

THE BENEFITS OF SELF-REGULATION

Since 1996 the UK model of self-regulation has been proven to be extremely successful, whereas attempts elsewhere to go down the route of prosecuting ISPs or introducing new legislation have failed. This is illustrated by the prosecution in 1996 of Felix Somm, the former head of the German division of CompuServe, for wilful distribution of pornographic content under Section 184 of the German Penal Code. This case, in which even the prosecution argued for an acquittal and appealed the original guilty verdict, shows the need for a collaborative rather than confrontational approach to criminal use of the net.

In a commentary on the appeal proceedings, in which Somm was acquitted, Professor Ulrich Sieber observes that:

> As a result, the acquittal of Felix Somm means not only the rehabilitation of an innocent citizen who was made a national hostage in the international Internet. Rather the case must also be seen as an opportunity to abandon absurd alibi solutions and replace the unnecessary confrontation between law enforcement and the Internet industry with an effective co-operation. The solutions of the next millennium demand inter-nationalisation, co-operation and an increased responsibility on the part of the citizen. The point in issue – reduced to its basics – is the need to develop a new culture of responsibility for the new Millennium.'
>
> (Commentary on the Conclusion of Proceedings in the Compuserve Case, www.digital-law.net/somm/commentary)

In considering the practical outworkings of this 'new culture of responsibility', Professor Sieber goes on to recommend that:

> An effective fight against Internet crime also requires – as opposed to the hitherto confrontation – a better co-operation between law enforcement authorities and the Internet industry. The law enforcement agencies and the Internet industry must recognize their common interest in the prevention of illegal contents in the Internet. They are natural allies, who can only experience success together, in tandem: law enforcement

agencies require the co-operation of the Internet industry particularly with respect to the speedy detection of criminal offenders in the international context; the Internet industry is dependent in many instances on the monopoly on power of law enforcement authorities for the prevention and detection of crime on the Internet.

The self-regulatory approach in the UK is based on this collaborative principle – indeed it should more properly be called 'co-regulation' – and through the work of the Internet Watch Foundation it follows three main strands: a notice and takedown procedure for illegal content, labelling and filtering for material which may be considered to be potentially harmful and education and awareness initiatives to tackle online safety issues.

The notice and takedown system works on the principle that if ISPs are provided with actual knowledge of illegal content on their servers, they can then remove it. The IWF hotline therefore receives reports from the public, accesses and assesses the reported material and advises ISPs if it is found to be potentially illegal.

Labelling and filtering acknowledges the fact that much of the internet content about which people may be concerned is not illegal and that therefore content providers have a right to upload it and users have a right to view it. However, parents and other carers in particular may consider it unsuitable for their children. Labelling works by means of an electronic label (metatag) being inserted in the HTML code for the web page which describes the content in objective terms. At the other end, users can set their own levels for 'acceptable' content using software available in the browser, so that any web page exceeding their own standards will be blocked.

The education and awareness remit commits IWF to providing information and advice on a range of internet content issues, particularly in terms of protecting young users from potentially harmful content and from inappropriate or even dangerous contact. IWF is working with the education, child welfare and sport sectors, as well as through the media, to get safety messages to young people.

In fulfilling these three functions, the IWF is sometimes accused of being self-appointed censors, on the basis that the internet should be allowed to function as a completely unregulated haven of free speech. The term 'censorship' in this context is entirely inappropriate.

CENSORSHIP

To take the area of illegal content first, law enforcement and censorship are not synonymous. As far as criminal content is concerned, the principle should and does apply that the internet is not a legal vacuum and therefore if something is illegal off-line it is also illegal online and should be dealt with accordingly – the internet should not be a safe haven for material or activity which has been deemed unacceptable in society generally. In most cases, the existing criminal law can and should simply be applied to the new medium, for example the IWF hotline primarily covers material which would be potentially illegal under the Protection of Children Act 1978 and in others the need for new laws has been identified, as in the case of the Regulation of Investigatory Powers Act 2000 (RIPA) or the proposed new legislation to deal with online 'grooming' of children for sexual purposes. So the process of ensuring that criminal content is removed and investigated is a legitimate part of law enforcement.

The notice and takedown mechanism on which the IWF hotline is based has been proven to work extremely successfully for child pornography and it is worth exploring in the context of other kinds of illegal or unlawful content. However, it is important to sound a note of caution. First, child pornography is unusual in that it is illegal even to possess, let alone to publish or distribute. This is not true of other categories of content and gives the ISP a liability which does not exist for other kinds of material. Second, the legal definition of child pornography is relatively clear, which makes the assessment of reported material equally relatively straightforward. This, too, is far less likely to be the case for many other kinds of content, particularly where civil matters such as defamation or copyright may be at issue. Each kind of content has to be considered separately and appropriate mechanisms found on a case-by-case basis.

A particular challenge for online policing has been highlighted by the terrorist attacks of 11 September 2001 and the potential use of the internet for communication between extremist groups. Although the atrocities are of an unprecedented scale, it seems appropriate to take a considered approach to how this challenge might best be met. It is essential that the rule of law be upheld nationally and internationally and therefore much dialogue is needed to find collaborative solutions which will offer the best chance of identifying effective measures to locate and take action against criminals while protecting the rights of legitimate users.

The charge of censorship is equally invalid in the area of potentially harmful content. Far from constituting an imposition of third-party censorship, the development of tools to help people make their own viewing decisions can maximize user choice without jeopardizing or restricting the right of free speech of those creating the material. Where potentially harmful content is concerned, the 'self' in self-regulation applies most appropriately to the individual user. It is not the role of government or non-governmental organizations to dictate what people can access, providing the material is legal (although commercial companies, as in all other areas of business, are entitled to make decisions about the kind of 'product' they wish to offer). But industry and self-regulatory bodies can play a valuable role in promoting the development and use of tools which will empower users to make their own decisions about what they do and don't want to view. Labelling is a mechanism that allows content providers to describe their material in an objective way and end users to select their personal viewing criteria for themselves or their children. This therefore preserves a balance between the right to freedom of expression at the one end and the right to freedom of choice at the other.

Education and awareness are essential tools for empowering users to control their own use of the internet, and that of children in their care, in order to help them exploit the positive potential of the net for communication and as a content resource. The acknowledgement of the fact that the activities of a minority of users may present potential hazards to some other, perhaps more vulnerable, users, is in my view an affirmation, not a denial, of the vast benefits of the internet to adults and children alike. We must recognize the genuine need to protect children online and, in the same way that we would not allow our children to wander unsupervised and indiscriminately around a large city, so we should take similar precautions in the online environment. There are simple 'stranger danger' rules with which most children would be familiar off-line, and these can generally be applied to their internet use.

At the same time it is important that the risks are seen in proportion and are not sensationalized; parts of the tabloid press have been quick to exaggerate the dangers, and this can be extremely detrimental to the process of formulating proportionate and effective solutions. There has been an increase in the number of children being contacted inappropriately through the net, but it is important to bear in mind that the risk of abuse is still far greater in the off-line world, particularly from family, friends and other trusted adults.

One issue which sometimes causes concern is that of poor quality information or even deliberate misinformation. Unlike other media, the internet has no editor to check that information is accurate, that it is not defamatory and that it does not breach someone else's intellectual property. A positive outcome of the lack of editorial oversight may well be that adults and children alike learn to develop enhanced critical skills of their own and certainly any educational programs involving the net should include this aspect as a core issue. The downside is the potential for dangerous content, such as misleading medical information or defamatory material. As for other abuses, it is essential that the criminal and civil law is adequate to deal with online content, that self-regulatory mechanisms are in place to make an informed response to such problems and that as far as possible users are given the tools to protect themselves.

The potential value of empowering and equipping individual consumers and their families was recognized by Professor Sieber in his commentary on the Compuserve case: 'Furthermore, the users of the Internet must also take on more personal responsibility. Within the domain of material harmful to minors, parents should become more pro-active, e.g. by educating their children of the dangers and using appropriate software products.'

To return to the issue of collaboration, at the time of IWF's inception in 1996 there was a potentially confrontational situation between law enforcement and the internet industry. By agreeing the principles on which the IWF is founded – that the law applies online as it does off-line, and that ISPs cannot reasonably be expected to monitor the contents of their servers – the basis was established for ongoing dialogue. This has been evidenced by the structure of the IWF itself, with a board which brings together industry representatives with nominees from a wide range of different sectors and by initiatives such as the Internet Crime Forum and the Home Office Task Force for Child Protection on the Internet, both of which illustrate the value of bringing a wide range of expertise to bear on the complex issues involved. The evolution of legislation such as RIPA also acknowledges the role that the internet industry has to play, in that it involves representatives from that industry in a technical advisory role and in the drafting of codes of practice to support the statutory legislative framework.

Internationally, too, the importance of cooperation is being increasingly widely recognized. Notwithstanding problems of differences in legislation and culture, as well as difficulties about issues such as jurisdiction and dual criminality, there is a high degree of collaboration between countries at governmental, police, industry and NGO levels. One example is the INHOPE (Internet Hotline Providers in Europe) Association which coordinates and promotes the work of hotlines across Europe and beyond through the exchange of reports and expertise and the development of best practice. Another is the Internet Content Rating Association, which is developing labelling and filtering systems and other technologies which will enable individual users to set their own standards for internet access according to their own personal, societal and cultural criteria.

The Information Society Directorate of the European Commission has devoted considerable resources to establishing and supporting

'Safer Internet' initiatives such as INHOPE and ICRA and is particularly committed to promoting international cooperation and encouraging 'European Added Value'. The EU's support for such programmes illustrates the fact that, far from being perceived as a menace to society, the benefits of the net are so great that all barriers to widespread usage must be tackled. The UK Government, too, is committed to promoting internet take-up in general and e-commerce in particular, as well as seeking to make the net as safe as possible for young users.

 **CONCLUSION**

The benefits of the internet, such as real time instant dialogue, the ability to be relatively anonymous, at least to other users and the potential for global one-to-many communication, inevitably have a flipside; the same technologies are available to the 'bad guys' as well as to the 'good guys'. Governments need to achieve a balance between encouraging the evolution and positive exploitation of this amazing tool and identifying the mechanisms to ensure that it is a safe environment for children and adults alike. But to portray the internet itself as a menace to society, would be tantamount to shooting the messenger. And with such a powerful messenger as this technology, society would ultimately be the loser.

NB While the views expressed in this essay have clearly been informed by my work with the Internet Watch Foundation, I would stress that they are my personal views and do not necessarily represent IWF policy.

Essay Four

INFORMATION TECHNOLOGY: PROSPECTS AND BARRIERS

Helene Guldberg and Sandy Starr

We seem today to have an intense love–hate relationship with information technology (IT). On the one hand, we endow IT with powers for all kinds of good; on the other, we fear its effects on our lives. Rather than seeing IT as a tool that can be used for good or for bad, IT tends to be seen as a force with the potential of governing every aspect of our lives – determining how we live, love, work and play.

Discussions about the internet following the terrorist attacks on the USA on 11 September 2001 provide a particularly striking snapshot of the way in which IT is perceived today. On the one hand, the internet was applauded for providing a 'bridge of communication' between loved ones across the world; e-mail and instant messaging worked when telephone networks in New York and Washington DC did not. Many observers eulogized the therapeutic and cathartic character of the internet, which was said to provide a 'convenient conduit for Americans to vent fears, frustrations and anger to a virtual community' (*CNET News.com*, 13 September 2001).

But commentators also lambasted the internet (often in the same breath) for being a 'haven of hate' – giving voice to anti-Arab and anti-Muslim hate-filled tirades – and for serving as a fertile breeding ground for hoaxers and spammers spreading misinformation online.

The Coalition Against Unsolicited Commercial Email (CAUCE) warned that many con artists were concocting online fraud, cynically profiting from our sympathies for the victims.

Even worse, some suggested that the internet should shoulder the blame for allowing the attack to happen in the first place. After all, the terrorists did use the internet when planning their bloody mission. They may have used the humble telephone as well, but strangely there was little discussion about the dangers of *that* technology.

The internet was condemned for facilitating terrorist organization, but at the same time welcomed as a new opportunity for state surveillance. The new technologies were simultaneously hailed for allowing survivors to communicate more easily and slammed for allowing terrorists to communicate too freely. The web was presented both as a goldmine of information and a minefield of misinformation.

Our love–hate relationship with the new technologies means that just as we fear the power of these technologies, we also look for technological solutions to almost every problem we face. An article in *USA Today* following the terrorist attacks proposed that 'the next time a terrorist aims a plane at a building, maybe the plane should say no. From computers that could steer jets away from skyscrapers to face-recognition devices used to spot card counters in casinos, technology could provide ways to make the skies safer' (1 October 2001).

This ambiguous and often polarized attitude toward the new technologies is not new. The technophile gurus of the dotcom era hailed the power of the internet to change everything. Industry pundits predicted that IT would revolutionize every aspect of our lives; getting rid of the need for everything from travel to supermarkets and transforming everything from workplaces and communities to the way we manage relationships. In his book *e-topia* (The MIT Press, 2001), architectural academic William J. Mitchell celebrates the transformation of the city into 'a whole new urban infrastructure ... long live the new, network-mediated metropolis of the digital electronic era.'

But just as IT has been presented as a solution to every major social problem, from global poverty and electoral disenchantment to educational standards and social inclusion, so it is seen as a hotbed of depravity and evil, playing a sinister role in spreading everything from paedophilia and pornography to childhood obesity and teenage suicide. As Andrew Calcutt points out: 'Cyberspace is often talked about as the latest in a long line of places where humanity is destined to screw up' (*White Noise: An A–Z of the Contradictions in Cyberculture*, 1999).

Today, the most immediate barrier to the deployment of IT is this negative attitude toward the new technologies. There has been a dramatic shift in attitudes to the internet from the dotcom frenzy of 1999 and early 2000 to the dotcom gloom of today, accompanied by creeping regulation and incursions on the autonomy of internet users. We are in danger of squandering IT's potential.

TO BOLDLY GO ....

It should be recognized that from a technical perspective IT – which is the fusion of two existing technologies: computing and communications – represents the most important technology of the end of the twentieth century. As Phil Mullan, CEO of the internet services company Cybercafé Ltd argues: 'Even if there is no basis to the claim that it has moved us beyond the industrial age, IT stands alongside the telegraph, the internal combustion engine, electricity, air travel and nuclear energy as a society-transforming technological breakthrough' (*spiked*, 12 February 2001).

IT could be a significant new tool for economic and social advance. The enhanced ability to store, process and transmit information has the potential to transform a vast array of business processes. But it takes time for the potential gains of a technology to be both recognized and realized. Take the process of electrification: although Michael Faraday made the initial discovery of electromagnetic induction in 1831, the first power stations did not open in Britain and New York until 1881. It then took over two decades for electricity to reach five per cent of households and businesses. By the late 1930s it reached half of households and three quarters of businesses. In other words, it took almost half a century for the more generalized deployment of electricity in Anglo-American industry after the first commercial electricity generator was established.

It is only reasonable to expect a prolonged gestation period for the application of the new information and communications technologies. But even this model of innovation requires long-term thinking and planning. And today's political and economic culture is inimical to the long-term investment in time and resources needed for IT to develop. The web has so far been used for its most 'quick fix' opportunities, as pointed out by Richard Lester, professor of nuclear engineering at MIT, who says that 'cost-cutting seems a more common reason for IT investment than expansion' (*The Productive Edge: A New Strategy for Economic Growth*, 2000). Today's negative perception of IT is the barrier to its deployment.

Following the big fall in the NASDAQ index of technology stocks in September/October 2000, Alan Skrainka, chief market strategist for the St Louis-based brokerage Edward Jones, told the *Financial Times* that 'the pendulum has swung from tremendous enthusiasm over technology and anything related to the internet to an environment of fear and concern' (13 October 2000). The extent of the gloom reveals the superficial and frivolous nature of the earlier hype. It also reveals the deep culture of pessimism and constraint that shapes the way we view the world today.

Is this pessimism warranted? Why did the 'internet bubble' collapse? Many blame the decadence, inexperience and incompetence of the new entrepreneurs. The BBC's internet and business correspondent Rory Cellan-Jones recounts in painful detail how, when the much-hyped boo.com launched its website, the website took eight minutes to load and had 396 technical bugs, while its workers (according to an ex-employee) 'were out every lunchtime getting shit-faced' (*dot.bomb: The Rise and Fall of Dotcom Britain*, 2001).

But the first thing that should be pointed out is that there was never really an 'internet' bubble in the first place; it was a 'stock market' or 'financial' bubble. The fall cannot be explained by the shortcomings of the new entrepreneuring nor by changes in global economic prospects. There were signs of a slight slowdown in economic growth in the USA prior to the collapse of the NASDAQ, but such a small fluctuation can hardly explain a fall of over 60 per cent since NASDAQ's peak in March 2000.

Rather, the stock market bubble contained the seeds of its own destruction. The financial markets became convinced by the hype that the dotcom sector offered easy short-term gains. This belief was strengthened by the fact that as investment went into the dotcoms the companies would often improve their financial performance – on paper if not in reality.

Therefore the bursting of the bubble can be explained by a huge gulf that opened up between the real economy – which includes companies involved in manufacturing and transport – and the financial economy. 'As the tempo of economic activity has slowed, many companies have found it more profitable to play the financial markets than invest in the productive economy,' explains financial commentator Daniel Ben-Ami (*spiked*, 20 March 2001).

This separation creates the potential for high financial volatility at the same time as sluggish economic growth. With such a large amount of 'liquidity', or surplus capital, circulating in the world's financial markets, there is always the possibility of financial bubbles developing rapidly. A small change in the outlook for a company or an economy can lead to grossly disproportionate changes in the financial markets.

There was no single event that pushed the markets down from their peak in March 2000; there was simply a growing anxiety that the markets were overinflated. Then over the following year a succession of disappointing results from companies such as Amazon, Cisco, Dell, Intel and Yahoo! created a cascade effect. The same markets that were enormously hyped became the subject of deep pessimism.

This separation between the real and financial economy has existed in general terms since the early 1970s. The new element over the past few years has been the increasing obsession with technology stocks and the 'new economy'. It is changing perceptions of the technology, media and telecoms (TMT) sector that have provided much of the impetus for stock market movements in Europe and North America in recent years.

Essentially, the TMT sector was driven up to its all-time high in March 2000 by overinflated expectations of its prospects. With the rest of the economy sluggish, it is not surprising that the financial markets became overexcited about one of the few areas of genuine economic dynamism. Much of the surplus capital circulating the world found its way into technology stocks, particularly in the USA.

But when the dotcom sector crashed, the vehement cynicism that followed was if anything worse than the blind optimism that had preceded it. As Rory Cellan-Jones rightly points out, those who 'smirked as lastminute.com's share price sank', who 'chortled as boo.com collapsed amid revelations of extravagance and management incompetence', did not make any contribution to the lessons learned from the period. And Cellan-Jones rightly derides the hypocrisy of 'pious talk from venture capitalists and analysts of how regrettable the "get rich quick" attitudes of so many firms had been', given that 'a few months earlier those same people had been trampling each other in the rush to cash in' (*dot.bomb: The Rise and Fall of Dotcom Britain*, 2001).

The online retailer Value America was the USA's first major dot.bomb, folding on the penultimate day of the twentieth century with what was then the biggest internet layoff of all time and kicking off the dotcom backlash. UK dotcoms, which had barely had time to catch up with the heady success of their US counterparts, suffered the same fate only five months later.

Unfortunately, naive speculators were not the only ones to suffer as a result of the rise and fall of the technology sector. These developments threaten to have a damaging impact on the prospects for what could be one of the most exciting areas for economic growth. The fall of technology stocks has helped create a mood of extreme cynicism about the new economy. Many of the same people who were once overhyping the sector are now wary of its potential. Those firms that could make genuine technological breakthroughs are likely to find it increasingly difficult to get funding. Projects which might have received an injection of capital may not do so today.

The extreme negative reaction – both within the internet world and among its erstwhile fans – is as unjustified as the earlier dotcom hype was inflated. A 'hold back, let's not risk the consequences', precautionary principle prevails and continues to be a huge handicap holding back the prolonged creative and experimental path required to make the most of our new technologies.

At the Don't Blow IT conference produced by *spiked* in London in September 2001, Phil Mullan usefully summed up our current predicament:

> I think we have had the worst of both worlds when it comes to failing to utilise these new technologies... The party phase – that frenzy of 1999 and early 2000 – was more cosmetic than real. Yet the hangover of negativity today draws its sustenance from the belief that there was genuine excess of resources going into the technology. The task we have as advocates of new technologies is less to find a cure for the hangover, but to get people to see that the party never really happened and that the real party needs to get underway.

AUTONOMY ONLINE

Although the economy has not yet had its internet party, the regulatory authorities, who initially saw the internet as a threat, have long been revelling in the opportunities that the internet gives them to erode the autonomy of users.

There is an interesting paradox at the heart of internet culture. On the one hand, the infrastructure of the internet was developed from the late 1950s under the auspices of the American state, as a command and control instrument during the cold war. Andrew Calcutt points out that 'the Information Age was first talked about nearly 40 years ago, at which time it was envisaged as the apogee of state-sponsored centralisation' (*White Noise: An A–Z of the Contradictions in Cyberculture*, 1999).

On the other hand, the early non-military adopters of the internet were from the techno-literate counterculture and believed that the medium would undermine state authority. These rebels saw the internet in what John Seely Brown and Paul Duguid call '6-D vision', as representing 'demassification', 'decentralisation', 'denationalisation', 'despacialisation', 'disintermediation' and 'disaggregation' (*The Social Life of Information*, 2000).

In the early 1990s the internet was also a place where the geeks ruled. Michele Tepper describes in her essay 'Usenet communities and the cultural politics of information' how cliques gathered online and codified their communication to one another through techniques such as 'trolling' (egging somebody on so as to expose them before the rest of the clique as ignorant of the ground rules) (in *Internet Culture*, D. Porter (ed.), 1997). These habits made early adopters feel as though they ruled the roost.

The icon of the empowered internet geek persists today, in the subculture of computer hackers and crackers and in the open source movement that developed the operating system Linux. These people make the internet seem like a world of plucky Davids and blundering Goliaths. But the state, the same state which funded the early development of the internet, has by no means retreated from it. The response to the September 11 terrorist attacks on the USA, a response which included sweeping new powers of US state surveillance, highlighted the fact that the internet has long since been a place where free expression and privacy are under threat.

Speaking on BBC Radio 4's *Today* programme, UK foreign secretary Jack Straw blamed the September 11 attacks upon 'a two-dimensional view of civil liberties' on the part of those who had defended liberty online and he claimed that such people 'will now recognize they were naïve in retrospect' (28 September 2001). At the risk of disappointing Jack Straw, far from 'recognizing that we were naïve in retrospect', those of us concerned with civil liberties recognize the threats to free speech and privacy that continue to handicap the internet's development.

FREE SPEECH UNDER THREAT

In his essay 'A brief history of censorship' Christopher Hunter identifies four stages that have been common to most forms of censorship throughout history: categorizsation, listing, word filtering and access and distribution control (in *Filters and Freedoms 2.0: Free Speech Perspectives on Internet Content Controls*, Electronic

Privacy Information Center (ed.), 2001). The internet has been no exception to this schema and currently looks to be entering the final phase: access and distribution control.

Since the four stages described by Hunter are made possible only by the technical means of reproducing material, censorship laws have traditionally been created that rely for their effectiveness on the limitations of the publishing technology of the era. The decentralized internet has proven as frustrating to modern censorship laws as the printing press was to church-dictated lists of banned books. Internet censorship has assumed two particular forms in response to this technical challenge, one old and one new.

The older form of censorship is one of moralism, where certain kinds of content are deemed beyond the pale, so abhorrent that they become special justifications for censorship. This moralism profits from the particular moral panics of the day and takes every opportunity to blur the distinction between thoughts, words and images, on the one hand, and actions on the other.

Take the lawsuit initiated against the US internet portal and search directory Yahoo! in France in April 2001, for hosting auctions of Nazi memorabilia. This case was made possible by an article of the French criminal code which prohibits the sale or display of any item that incites racism. This law takes a leap from an item with racist connotations (which understandably causes offence) to actual racist violence and claims that there is a direct causal relationship between them.

This sleight of hand is not unusual. US free speech campaigner Marjorie Heins instructs us in a standard justification of internet censorship on moral grounds, summarizing a supposedly liberal 1997 speech by the Democrat Al Gore as 'formulaic genuflections to the First Amendment with calls for blocking minors' access to "offensive speech" that he analogized to various physical dangers – live electrical outlets in one metaphor; poisons in the medicine chest a moment later' (*Not in Front of the Children: 'Indecency', Censorship, and the Innocence of Youth*, 2001). Heins traces child-based justification for censorship back to the key case of Regina *v.* Hicklin in 1868, which established the 'Hicklin test' for legal assessment of controversial material. This test holds that material should be outlawed if it might harm its most vulnerable consumers, setting a lowest common denominator standard for free speech. The test has since been moderated as a legal standard, but it continues to epitomize the patronizing spirit of all internet censorship that takes place in the name of children.

Authorities on both sides of the Atlantic have profited from paedophile scares to establish legal precedents for conflating the actions of child sex offenders with artificially created images distributed online. The UK's Criminal Justice and Public Order Act 1994 revised a previous piece of legislation outlawing child pornography, the Protection of Children Act 1978, altering every reference to an 'indecent photograph' to refer instead to an 'indecent photograph [or pseudo-photograph]'. Two years later in the USA, the Child Pornography Prevention Act 1996 was signed into law by Bill Clinton, again revising a previous piece of legislation that dealt with child pornography. A depiction of a minor was now illegal where 'such visual depiction has been created, adapted or modified to appear that an "identifiable minor" is engaging in sexually explicit conduct' ('identifiable minor' meaning identifiable as *being* a minor, not necessarily identifiable as a particular minor).

Legal developments such as these paved the way for internet censorship, by falsely characterizing the internet as a medium for actions, rather than what it really is: a medium for information. While the internet is a (sometimes disturbing) document of the extremes of human interests, it tells us nothing about the extent of human actions. The American Civil Liberties Union (ACLU) rightly challenged the Child Pornography Prevention Act 1996 on the grounds that 'there is a real difference between touching children and touching computer keys to create images' (brief of ACLU *et al.*, as Amici Curiae, Free Speech Coalition *v.* Reno, 1997).

Moralism aside, the newer form of censorship that has flourished online is a particular kind of self-censorship, where regulation devolves from accountable arms of the state to unaccountable para-state, international or industry bodies, a defence mechanism where censorship mimics the decentralization that frustrates the efforts of state censors to regulate the internet in the first place.

This devolved censorship has been defended as being democratic, on the grounds that rather than being dictated officially by the state, an institution or person opts for it voluntarily. So the Internet Watch Foundation (IWF), responsible for the vast majority of removals of internet content in the UK, has powers that according to some studies outstrip those of the British state 400-fold (C. Ellison, 'Oppression net', *Economic Affairs*, March 2000). But the IWF has responded to the charge of being anti-democratic by pointing out that internet service providers (ISPs) elect to remove material on its instructions and are not legally mandated to do so. This despite the fact that material has been removed from the internet without the consent or knowledge of the internet user.

So much for democracy. But what of those instances where the individual internet user *is* the one who opts for censorship of the material they view – for instance, if the user buys a censorware product? The ACLU's landmark 1997 pamphlet *Farenheit 451.2: Is Cyberspace Burning?* (reproduced in *Filters and Freedoms 2.0*) attempted to address this question, describing how makers of filtering software could collude with web browser manufacturers, search engines and government-backed 'self-rating' schemes to create a chilling effect on free speech online, even in the absence of laws directly censoring the internet.

However, the ACLU conceded in its pamphlet that 'user-based blocking programs ... are far preferable to any statute that imposes criminal penalties on free speech.' Does this mean that self-censorship by the individual internet user is acceptable? Not according to the American legal theorist Lawrence Lessig, who argues convincingly in his book *Code and Other Laws of Cyberspace* (1998) that filtering opted for by internet users is dangerously 'non-transparent'. 'If there is speech the government has interest in controlling', says Lessig, 'let that control be obvious to the users. Only when regulation is transparent is a political response possible.'

In other words, the danger of censorship mediated through unaccountable institutions or even through the individual internet user is that it takes decisions with political consequences and removes them from the sphere of political debate. If it is true, as John Fitzpatrick of the Kent Law Clinic argued at the Don't Blow IT conference (citing Georg Lukács) that 'freedom is an *activity*', then any form of self-censorship is particularly harmful to freedom of expression, because it precludes the activity of freedom from even getting underway.

That is why self-censorship is so insidious and self-propagating: it begets further self-censorship. In the Yahoo!/Nazi memorabilia case, for instance, it went largely unnoticed that Yahoo! had opened itself up to the French lawsuit through a prior instance of self-censorship. In September 1999 Yahoo! had been commended before a US Senate judiciary committee for removing over 70 websites from its listings that were judged to be run by 'hate groups'.

This (entirely voluntary) gesture was great PR for Yahoo!, but it also implied that Yahoo! bore responsibility for content that it hosted, a view that the plaintiffs in the subsequent Nazi memorabilia case evidently shared. It is regrettable that in January 2001, even as Yahoo! contested the French court's ruling in the Nazi memorabilia case, the company announced that it would (again voluntarily) begin self-regulating the auction of Nazi-related items on its website, using special software. This second PR-friendly stunt merely added moral weight to the initial case against Yahoo! and thus propagated the cycle of self-censorship still further.

PRIVACY REDEFINED

Privacy has traditionally been defended in the context of civil liberties, as part an individual's protection against intrusion by the state. The right to privacy is implicit in the Fourth Amendment of the US Constitution, which guarantees 'the right of the people to be secure in their persons, houses, papers, and effects, against unreasonable searches and seizures' and the Fifth Amendment, which protects against 'private property' being 'taken for public use'.

Privacy from market agents has traditionally been a distinct (and lesser) concern that falls under the auspices of trading standards enforced by the state. Since the development of information technology has enabled electronic credit card transactions and customer profiling databases, privacy in the trading standards sense has assumed increased prominence in public debate. With the explosion of internet use in the 1990s and the concurrent popularity of the notion of 'consumer rights', privacy from companies actually eclipsed privacy from the state as a public concern.

It is true that internet technology raises specific difficulties in protecting privacy. Everything we do online, whether buying goods, reading articles or simply browsing, creates what the American legal commentator Jeffrey Rosen describes in his book *The Unwanted Gaze: The Destruction of Privacy in America* (2000), as 'electronic footprints' that reveal details of our personal habits, interests and even thoughts.

What is the private or public status of these 'electronic footprints' and who should be entitled to possess or peruse them? With other forms of communication, generic information (addresses on envelopes, telephone numbers) can be separated from content (letters, telephone conversations). To have access to the former is not necessarily to have access to the latter, which is why one enjoys a degree of privacy from the post office and from the telephone company. But on the internet such distinctions break down and even our reading habits are exposed to public view.

Does online privacy matter? It has been suggested that since the online protection of privacy from companies, employers and the state is so difficult, we should either resign ourselves to treating the internet as public space or alternatively all adopt elaborate encryption techniques. Sun Microsystems CEO, Scott McNealy, asked about internet privacy issues at a 1999 press conference, replied: 'You already have zero privacy. Get over it.' Jeffrey Rosen correctly dismisses this bleak view as 'a bovine surrender to technological determinism, putting us in the same category as citizens in the Soviet Union' (*The Unwanted Gaze: The Destruction of Privacy in America*, 2001).

Rosen is not alone in his convictions. Over the past decade, a plethora of lobby groups, privacy advocates and encryption/anonymity/pseudonymity technologies has emerged to protect privacy online. Because the most immediately obvious benefit of monitoring internet behaviour is to corporations that can collect and distribute people's demographic and purchasing information, the principal target of the privacy activists' concerns has been commercial bodies. This has been particularly true in the wake of online customer confidentiality gaffes by brands as big as Barclays and in the wake of Toysmart.com's much-vilified attempt to sell its customer databases after the company went bust.

The consequence of these concerns about company use of consumer information has been a shift in the understanding of what privacy means and why it is important. This has led to a rise in interventionist legislation by state and by international regulatory bodies. More than 300 privacy-related bills have recently been put forward for consideration in state legislatures across the USA. According to some surveys, existing EU data protection laws are currently violated by 80 per cent of European websites, leading Ben Vickers, the European Commissioner for Enterprise and the Information Society, to warn ominously of 'changes in EU law that ... will close the loopholes of new technologies' (*Wall Street Journal*, 15 February 2001).

Fear of the misuse of consumer information by market agents is understandable and trading standards provide important protections against such misuse. But the trouble with entrusting the protection of privacy to the state and asking the state to set privacy standards through legislation is that the state subsequently becomes empowered to violate privacy whenever such violation is seen to be in the public good.

In the UK, for instance, the Regulation of Investigatory Powers (RIP) Act 2000 grants the state authorities unprecedented powers to monitor electronic communications and makes it mandatory, if one is requested by police or security services to provide one's computer passwords or encryption keys, to do so.

One can scarcely imagine a more direct or a more symbolic intrusion upon privacy by the state than the RIP Act. And yet while consumer privacy concerns maintained their prominence throughout 2000, the RIP Act was met with near silence. The UK's Foundation for Information Policy Research (FIPR) campaigned against the Act's introduction, and was instrumental in modifications to the Act that reduced its impact upon privacy. But as FIPR director Caspar Bowden told the audience at Don't Blow IT, his attempts to interest the public and the media in the consequences of the Act were an uphill struggle.

The discussion about privacy was not posed in terms of civil liberties that year, but it was very much the following year. The September 2001 terrorist attacks on the USA prompted a reversion of privacy concerns from the threat of companies to the threat of the state. Ironically, the state's role in protecting privacy from commercial intrusion had already strengthened its hand. US attorney general

John Ashcroft ushered in expanded powers of state surveillance, on- and off-line, to prevent future terrorism. While Jack Straw complained that defenders of civil liberties had aided the terrorists by resisting the RIP Act, Ashcroft seemed to be looking to the RIP Act as a model for US legislation such as the Uniting and Strengthening America by Providing Appropraite Tools Required to Intercept and Obstruct Terrorism (USA PATRIOT) Act.

If we can learn anything from developments in the internet privacy debate, it is that privacy can be both a progressive and a destructive cause. In its most general terms, the private sphere is a laudable thing to fight for. It is a place where one can indulge in what philosopher John Stuart Mill, in his famous 1859 essay *On Liberty*, called 'experiments of living' where 'individuality should assert itself', free from the scrutiny of the outside world.

This private sphere needs to be defended against any party who would intrude on it, whether that party be the state, a company that one does business with or one's own employers, whose latitude to snoop upon you, by electronic or any other means, should be kept to an absolute minimum. But once this private sphere has been established and defended, its proper use is as a bedrock for public activity and accountability to others. If it is defended in isolation and for its own sake, privacy becomes a pretext for lack of social engagement. Online, privacy can become a pretext for only communicating anonymously or for restricting the development of *any* commercial technology that involves the submission of private details to a company. So Microsoft's Hailstorm initiative, a potentially revolutionary technology that uses personal information to integrate web services, was shot down for its privacy implications before it was properly assessed for its technological potential.

In order for the internet to achieve its full potential as a communications medium, it is necessary to both defend online privacy vigilantly against intrusion *and* be willing to submit personal information online where appropriate. These two preconditions are not contradictory but complementary. The internet should be neither a snooping ground for the state nor an atomized collection of anonymous individuals (and it is presently in danger of becoming both), but rather a place where members of society can communicate and collaborate to progressive ends.

 **CONCLUSION**

Technological determinism – the tendency to ascribe autonomous power to IT, to consider it the independent variable in social change – is at the root of most obstacles to realizing the full potential of the new technologies.

Two recently published books, *The Internet Galaxy* by Manuel Castells and *Ecosystem* by Thomas Power and George Jerjian, embrace metaphors for the internet (one astronomical, one ecological) that imply that the medium has a complex life of its own and that the people and businesses who use it are just cogs in the greater system, pieces of the whole. This thinking is dangerous. In the case of technophiles, it presents IT as generating social benefit of its own accord, independent of the work of people and thus sets the medium up for a fall. In the case of technophobes, it characterizes the internet as the active agent in any harm or tragedy to which it is related, from the pettiest crime to the mass terrorism of September 2001, prompting irrational panic and inviting draconian regulation.

In the case of business, such an approach leads people to make a virtue of spontaneous developments in the internet and its application at the expense of long-term strategy, a tendency which helps explain how the crash of the (relatively) tiny dotcom sector and associated financial markets could have a domino effect upon the economy as a whole.

In the case of privacy, this thinking characterizes technology itself as the intrusive enemy and therefore puts a brake on technological developments. Moreover, by blurring the state apparatus with the marketplace, it obscures the true identity of those who would actively use the technology to erode our privacy.

In the case of free speech, it reduces political debate to the level of technical quibbling. In legal disputes concerning internet content, plaintiffs hold defendants responsible for internet content, defendants protest that they cannot technically be responsible for that content and IT buffs are brought in to resolve the matter.

Lazy defenders of free speech adopt this thinking to assume that the medium of the internet is intrinsically resistant to regulation, but this belief is quixotic; wherever regulation *is* technically possible, it tends to be pursued. A passive attitude toward free speech online results in a situation where free speech is defined by technical default, rather than by principled conviction. As Lawrence Lessig explains, it a common fallacy that 'the net has a nature, and that nature is liberty.' Rather, 'the possible architectures of cyberspace are many... Some architectures make behaviour more regulable; other architectures make behaviour less regulable' (*Code and Other Laws of Cyberspace*, 1998).

The internet should be seen not as having a benevolent or malevolent life of its own, but as being a tool of enormous potential, with which we can achieve concrete human goals.

Let's not blow IT.

AFTERWORD
Dolan Cummings

The internet is an exciting technology that can improve the lives of individuals and can be good for society in various ways. That much seems to be uncontroversial. Nobody in this book has argued that the internet is a bad thing or that it ought to be banned or suppressed. Good. Far more interesting is the controversy over how to make the most of the internet. The answer to this question seems to rest in part on one's understanding of how exactly the internet is changing society.

There is a degree of consensus that the internet is not as revolutionary as it is often claimed. The exaggerated claims about the internet are symptomatic of a peculiarly extreme brand of technological determinism doing the rounds recently. This type of worldview, in which technology rather than people are the main movers, tells us a lot more about the meagre expectations society holds of what people can or cannot do than it does about any unusual features of the latest technologies. It is different expectations of how people might use the internet that has emerged as the main source of contention in the essays in this book.

HOW TO MAKE THE MOST OF THE INTERNET

AN ASPECT OF CONSUMPTION

It is important to recognize that the internet has emerged at a particular time and, in spite of its global character, in particular places. The internet's development is shaped by its historic and geographic roots as much as by factors intrinsic to the technology itself. For many, the most important influence on the internet is consumerism.

While the technology can be picked up and adapted for various different uses by different people in different places, in a society in which, as Peter Watts puts it, people 'render their lives meaningful through decisions of personal consumption', our understanding of the internet is likely to be dominated by its potential as a means of consumption. Undoubtedly, the internet is used extensively for shopping. People shop online not only for books and CDs, but also for clothes, holidays and even for cars and insurance policies.

A TECHNOLOGY WITH UNMET POTENTIAL

Helene Guldberg and Sandy Starr have rather higher expectations. Their website, *spiked*, is billed as a site for those who believe that 'the power of the internet could be used for something more than shopping and pseudo-sex'. For some, the internet could be used to develop the economy much more comprehensively and to improve everybody's standard of living. To achieve this, however, we cannot simply depend on the technology itself. Long-term planning based on a realistic assessment of the technology's potential is needed to make the most of the internet for business and for society in general.

A VALUABLE FORUM TO BE PROTECTED FROM ABUSE

Non-technical considerations also have to be given careful attention. Probably the most important of these is the question of free speech online. Again, there is a general consensus that the internet can be of huge benefit as a means of publishing and distributing ideas and information. The question of how and whether the medium should be regulated is more controversial.

Probably, some users are put off by the idea of a totally unregulated internet. In particular, the existence of illegal material, whether this is pornography or hate speech, can be frightening, especially for parents. Traditional censorship, however, is almost impossible. The kind of self-regulation encouraged by the Internet Watch Foundation is one alternative. Fundamentally, this approach is based on the existing law, which, it is argued, ought to be applied online just as it is applied off-line. In a sense, then, self-regulation is a technical matter, intended to normalize the internet, rather than a sinister form of censorship.

A SOURCE OF NEW FREEDOM

For some free speech campaigners, however, self-regulation is even less accountable than old-fashioned censorship. Devolving power from the state to self-appointed regulators and internet service providers makes censorship a technical matter, rather than one that is open to public scrutiny. Certainly, defenders of free speech have to address laws that apply off-line as well as on. Undoubtedly, though, it is much more difficult for the authorities to regulate the internet than printed publications. The question is whether we see this extra freedom as an advantage to be exploited and celebrate the anarchy of the internet or whether we try to bring the internet into line with other media.

78

DEBATING MATTERS

A similar tension exists in the discussion about intellectual property online. The internet has certainly revolutionized our ability to copy and distribute text and music. Legislation that was established to protect the livelihoods of authors and musicians at a time when publishing meant printing has proved almost impossible to police in the face of the new technology. The question posed by Chris Evans is whether we should endeavour to make the law work to protect intellectual property or whether we should take the opportunity to rethink the idea of copyright altogether.

Clearly, the internet offers challenges as well as opportunities, and plenty to argue about. Fortunately, it also serves as an excellent forum for debate and no doubt this discussion will continue online. The internet does not exist in a brave new world of its own, but it certainly exposes us more fully to the world we already have. How we respond to this technology depends more than anything on how we view that world and the people in it.

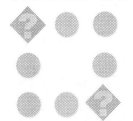

DEBATING MATTERS

Institute of Ideas
Expanding the Boundaries of Public Debate

If you have found this book interesting,
and agree that 'debating matters', you can
find out more about the Institute of Ideas
and our programme of live conferences and
debates by visiting our website
www.instituteofideas.com.
Alternatively you can email
info@instituteofideas.com
or call 020 7269 9220 to receive a full
programme of events and information about
joining the Institute of Ideas.

Other titles available in this series:

DEBATING MATTERS

Institute of Ideas
Expanding the Boundaries of Public Debate

SCIENCE:

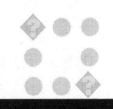

CAN WE TRUST THE EXPERTS?

Controversies surrounding a plethora of issues, from the MMR vaccine to mobile phones, from BSE to genetically-modified foods, have led many to ask how the public's faith in government advice can be restored. At the heart of the matter is the role of the expert and the question of whose opinion to trust.

In this book, prominent participants in the debate tell us their views:

- Bill Durodié, who researches risk and precaution at New College, Oxford University
- Dr Ian Gibson MP, Chairman of the Parliamentary Office of Science and Technology
- Dr Sue Mayer, Executive Director of Genewatch UK
- Dr Doug Parr, Chief Scientist for Greenpeace UK.

ART:

WHAT IS IT GOOD FOR?

Art seems to be more popular and fashionable today than ever before. At the same time, art is changing, and much contemporary work does not fit into the categories of the past. Is 'conceptual' work art at all? Should artists learn a traditional craft before their work is considered valuable? Can we learn to love art, or must we take it or leave it?

These questions and more are discussed by:

- David Lee, art critic and editor of *The Jackdaw*
- Ricardo P. Floodsky, editor of artrumour.com
- Andrew McIlroy, an international advisor on cultural policy
- Sacha Craddock, an art teacher and critic
- Pavel Buchler, Professor of Art and Design at Manchester Metropolitan University
- Aidan Campbell, art critic and author.

ALTERNATIVE MEDICINE:

SHOULD WE SWALLOW IT?

Complementary and Alternative Medicine (CAM) is an increasingly acceptable part of the repertory of healthcare professionals and is becoming more and more popular with the public. It seems that CAM has come of age – but should we swallow it?

Contributors to this book make the case for and against CAM:

- Michael Fitzpatrick, General Practitioner and author of *The Tyranny of Health*
- Bríd Hehir, nurse and regular contributor to the nursing press
- Sarah Cant, Senior Lecturer in Applied Social Sciences
- Anthony Campbell, Emeritus Consultant Physician at The Royal London Homeopathic Hospital
- Michael Fox, Chief Executive of the Foundation for Integrated Medicine.

COMPENSATION CRAZY:

DO WE BLAME AND CLAIM TOO MUCH?

Big compensation pay-outs make the headlines. New style 'claims centres' advertise for accident victims promising 'where there's blame, there's a claim'. Many commentators fear Britain is experiencing a US-style compensation craze. But what's wrong with holding employers and businesses to account? Or are we now too ready to reach for our lawyers and to find someone to blame when things go wrong?

These questions and more are discussed by:

- Ian Walker, personal injury litigator
- Tracey Brown, risk analyst
- John Peysner, Professor of civil litigation
- Daniel Lloyd, lawyer.

NATURE'S REVENGE?

HURRICANES, FLOODS AND CLIMATE CHANGE

Politicians and the media rarely miss the opportunity that hurricanes or extensive flooding provide to warn us of the potential dangers of global warming. This is nature's 'wake-up call' we are told and we must adjust our lifestyles.

This book brings together scientific experts and social commentators to debate whether we really are seeing 'nature's revenge':

- Dr Mike Hulme, Executive Director of the Tyndall Centre for Climate Change Research
- Julian Morris, Director of International Policy Network
- Professor Peter Sammonds, who researches natural hazards at University College London
- Charles Secret, Executive Director of Friends of the Earth.

DESIGNER BABIES:

WHERE SHOULD WE DRAW THE LINE?

Science fiction has been preoccupied with technologies to control the characteristics of our children since the publication of Aldous Huxley's *Brave New World*. Current arguments about 'designer babies' almost always demand that lines should be drawn and regulations tightened. But where should regulation stop and patient choice in the use of reproductive technology begin?

The following contributors set out their arguments:

• Juliet Tizzard, advocate for advances in reproductive medicine
• Professor John Harris, ethicist
• Veronica English and Ann Sommerville of the British Medical Association
• Josephine Quintavalle, pro-life spokesperson
• Agnes Fletcher, disability rights campaigner.

TEENAGE SEX:

Under New Labour, sex education is a big priority. New policies in this area are guaranteed to generate a furious debate. 'Pro-family' groups contend that young people are not given a clear message about right and wrong. Others argue there is still too little sex education. And some worry that all too often sex education stigmatizes sex. So what should schools teach children about sex?

Contrasting approaches to this topical and contentious question are debated by:

- Simon Blake, Director of the Sex Education Forum
- Peter Hitchens, a columnist for the *Mail on Sunday*
- Janine Jolly, health promotion specialist
- David J. Landry, of the US based Alan Guttmacher Institute
- Peter Tatchell, human rights activist
- Stuart Waiton, journalist and researcher.